COMMON-SENSE C
BY THE QUICK F

COMMON-SENSE COMPOST MAKING
By the Quick Return Method

by
MAYE E. BRUCE

with a foreword by
L. F. EASTERBROOK

new edition revised by
LADY EVE BALFOUR

FABER AND FABER LIMITED
3 Queen Square
London

*First published in 1946
by Faber & Faber Limited
3 Queen Square London WC1
New edition completely revised 1967
Reissued in this edition 1973
Reprinted 1975 and 1977
Printed in Great Britain by
Whitstable Litho Ltd., Whitstable, Kent
All rights reserved*

*ISBN 0 571 09990 4
(Faber Paperbacks)*

Copyright in this book was left by the author to The Soil Association to which all royalties are paid

CONDITIONS OF SALE

This book is sold subject to the condition that it shall not, by way of trade or otherwise, be lent, re-sold, hired out or otherwise circulated without the publisher's prior consent in any form of binding or cover other than that in which it is published and without a similar condition including this condition being imposed on the subsequent purchaser

© 1967 The Soil Association

Contents

EDITOR'S PREFACE TO THE REVISED EDITION	*page* 9
FOREWORD BY L. F. EASTERBROOK	13
1. INTRODUCTORY NOTE	17
2. THE STORY OF QUICK-RETURN COMPOST	22
3. THE HOW AND THE WHY OF THE HEAP	33
4. THE COMPOST AND THE GARDEN	51
5. EFFECT ON HUMAN HEALTH	60
6. THE ACTIVATOR	65
7. THE CONVICTION	74

Appendices

1. TABLE I. BUILDING THE GARDEN HEAP	77
TABLE II. BUILDING THE STRAW HEAP	77
TABLE III. MANURE TUBS	78
TABLE IV. LEAF HEAP	78
TABLE OF FAILURES, CAUSES AND REMEDIES	78
2. FORMULAE FOR HERBAL POWDER	79

Appendices

3. PRICE OF POWDER ACTIVATOR *page* 81
4. HERBS—AND WHERE TO FIND THEM 81
5. ALTERNATIVE PLANTS AND THEIR CONSTITUENTS 83
6. SOME USEFUL HINTS 83
7. PLAN FOR A SMALL BIN 85
8. PLAN FOR A MOVABLE BIN 87
9. TESTS AT THE HAUGHLEY RESEARCH FARM 89
10. THE SOIL ASSOCIATION 92
    BIBLIOGRAPHY 94
    INDEX 95

*'The Divinity within the Flower
is sufficient of itself'*

Editor's Preface to the 1967 Revised Edition

The late Miss Maye Bruce's innumerable disciples (of whom I am one) will welcome this re-issue of her admirable little book *Common-Sense Compost Making*, and I am convinced that the new generation of gardeners, especially home-gardeners, have only to read it to be equally delighted —even experienced compost gardeners will find in it a useful reference book. In the process of revising it, I discovered a number of useful tips which I had forgotten; I am sure others will find the same.

A few words are necessary about the revision. Since I started out with the intention of making as few alterations as possible, I was delighted, and somewhat astonished, (seeing that the last edition was published in 1946) to find how very little was out of date. I have deleted a few paragraphs which are not applicable to today's conditions; I have brought a few facts up to date, and altered a few tenses, where not to do so would have given misleading information, (for example, the prices I have quoted for the cost of the herbal activator are those ruling in 1967), and I have transferred some of the information given in a number of appendices to the main text. This has involved a slight rearrangement of the sequence of the latter, but readers familiar with the original version will find all the relevant matter still there.

I have not had to alter a word of the late Laurence Easter-

Editor's Preface to the Revised Edition

brook's introduction, and on re-reading it, I felt, once again, how often his writing is time-less. I have also deliberately left unaltered the author's occasional references to the period—phrases such as 'in these war-time days', because I find it historically most interesting to note how many of the trends, only beginning when the book was written, have developed as foretold—sometimes frighteningly so—as in the case of the increase in toxic sprays, and environmental pollution generally and sometimes encouragingly—as in the case of the spread of the demand for whole food. By contrast Ministries, particularly of Health, have hardly moved at all.

On page 63 Miss Bruce tells the following story: 'At the Anthroposophical farm in Holland, the produce was sent to customers direct, and by a specialized system of delivery and order. In time a certain family demurred at the extra price, and returned to the market stall. After some time they came back to their old allegiance, saying they had had to pay so much in doctors' bills since eating the market stuff, that it more than counterbalanced the higher prices paid for vital food!'

In view of the ever mounting costs of medical services, I find it somewhat depressing that successive Governments still remain determinedly and blindly unaware of such experiences—which could be multiplied up and down the country, and in many parts of the world.

Of course the answer, as in all social reforms, is that the public must demand before Governments will act, and, of course, before public demand can arise, public awareness must develop. The signs are rife that this process has begun.

Miss Bruce's prophetic voice; her enthusiasm, and her selfless giving of her inspiring personality, and the benefit of her personal experience, were potent factors during her lifetime in helping to awaken this awareness. The time is

Editor's Preface to the Revised Edition

exactly right for her voice to be heard again. May the new edition of *Common-Sense Compost Making* bring it into many new homes, along with the satisfaction that compost gardening brings.

One word of warning, however,—in her way, Maye Bruce was a priestess of her craft. There is no doubt that there is some intuitive factor involved in those who have 'green fingers'. I suspect the same thing can apply to compost making. I have never achieved quite the speed or perfection Maye describes in her book, and which habitually took place in her own compost heaps, as I have many times seen for myself. But let the beginner not be discouraged by this. The purpose of it all is the vigour, quality, disease resistance, and flavour of the vegetables grown with the compost—and the end result is the improved health of the consumer. These things will come about if the advice in this book is followed, even if absolute perfection in the compost itself does not.

EVE BALFOUR

Note to 1972 Edition

Very little further revision has been found necessary: information on the supply of Powder Activator (p. 81) has been revised and Appendix 10 on The Soil Association has been brought completely up to date.

Foreword
By L. F. Easterbrook

When Dr. Rudolf Steiner was pressed to lecture publicly on agriculture, he eventually agreed, but with reluctance. 'All right', he said, 'this is what I think. But for Heaven's sake experiment for yourselves.' That is precisely what Miss Bruce has done, as the fascinating story she tells in Chapter 2 reveals. With her usual modesty, she puts forward the method that she has developed as merely one of three, each of which may be suited to particular circumstances. Hers is especially suitable to gardens and allotments and to the increasing number who find adequate supplies of animal manure hard to come by. But Miss Bruce would be the last person to claim to be the sole repository of knowledge about compost making, and her readiness to recognize the claims of other systems is further proof of the spirit of disinterested public service in which she has undertaken this work. She seeks neither prestige nor financial reward.

Had I not been already convinced of this, I could not have taken the responsibility of giving a full description of her methods in a national newspaper. The result was staggering. Over 4,000 people wrote in the next few weeks to ask for further particulars. The fact was that it appealed to their common sense.

That seems to be the most remarkable thing about this business of fertilizing the soil by completing the circle of

Foreword

growth and returning to the earth organic matter that has served its immediate purpose. It is Nature's way, and although Nature is far less doctrinaire than many who fancy they can ignore her, and although she will permit liberties to be taken, yet she is inclined to be implacable when it comes to going against her first principles. The lack of health and the growing catalogue of diseases in plants, animals and men seems to me evidence of this, and in our hearts I believe we know it is true.

For when I first encountered the theories of Dr. Steiner and the methods of those who believe that only living things can produce life, I was frankly incredulous. That was some fifteen year ago (1930). But somewhere there lurked the uncomfortable feeling that 'there was something in it'. This led me to try it, rather tentatively, about ten years ago. 'But I'm not going all the way', I said to myself, 'one mustn't become a crank'. I found myself going further and further, however, and even when I have metaphorically shaken my fist at Dr. Steiner's photograph and said that anyhow nothing would make me believe that one, sooner or later I have had to make an equally metaphorical apology. So far as I can discover, this has been the experience of everyone who has set foot upon this path. Perhaps, therefore, this should have been headed 'Warning' and not 'Foreword'.

Today, after ten years' experience, all too literally at 'first hand' during the war years, I am completely satisfied with the result of Miss Bruce's system. Since the war, we have added poultry and rabbit manure to the vegetable waste, and it seems to have improved the compost. Our soil, inclined to be rather sticky on top of marl rock and unweathered greensand, has improved beyond all knowledge, and we can get on to it, and work it, for at least an additional six weeks during the year. The flavour of what we

Foreword

grow is at least noticeable enough to provoke invariable comment from visitors. It is true we pick the white butterflies off the cabbages, dust some of the young brassicas with derris and spray the roses with soft soap, but apart from that we use no spray or insecticide of any sort. Yet potato blight is unknown to us, likewise the other curses of the gardener, and if the peach trees suffer from the attentions of red spider, they have the vitality to throw off the effects. As regards other fruit, I wish I knew where we could buy apples to equal those that we grow, although no spray ever pumps lead and arsenic into them; and while the diseases of strawberries and raspberries have wrought such havoc that the national acreage has decreased by about 50 per cent, our main trouble is to restrain their exuberance in throwing out runners and suckers. I have never seen a sick strawberry plant or raspberry cane in the garden.

Our health has been good enough to make people ask us how we manage it. When I take my small boy who has eaten compost-grown vegetables all his life, to the dentist, the dentist asks what we have done to him to give him such an exceptional set of teeth. We hope to be even healthier now we have discovered where we can buy flour from organically-manured wheat.

Not the least of the blessings of this 'common-sense' gardening is to be free of the slavery of measuring out and administering endless doses of this or that dope to the square yard, making one feel more like a chemist's assistant than a gardener. Nature leaves wide margins for error, and will never quarrel over a few tons of compost to the acre one way or the other. Provided reasonable care is given in making the compost, it is as near to being a fool-proof system of manuring as anything can be. It makes the minimum demand upon intelligence and labour.

Foreword

All this, I am quite ready to agree, is only 'evidence' and not 'proof'. Those 'scientists' who so stoutly fight the losing battle of chemical manuring and are retiring from one position to another, will not even regard it as 'scientific' evidence. But with the rising tide of practical evidence that refutes their theories, the onus has come to be upon them to prove their contentions, and until they can produce proof that artificial manures will give me food of the quality, health and succulence of compost-grown food, and with such little trouble, I shall be content to keep to compost, save my money, and retain my well-being.

1945

CHAPTER 1

Introductory Note

Here is a curious fact. It has taken a World War to revive and strengthen the human love of the soil.

Throughout the ages, we find that work on and with the soil has meant fertility, health, prosperity; but as soon as man began to exploit it for gain, or neglect it from sloth, fertility ceased, the life departed from the earth, soil erosion followed, and vast tracts of land were invaded by sand and dust, with the result that once fertile country was turned into desert and dust bowl, and the process still goes on.

Nature is slow to retaliate, but terribly sure. The lesson may be learnt on every continent, either as the result of neglect in the long past, or from the concentrated and constant exploitation of a century. The first, neglect, is typified by the deserts of North Africa—once the granary of Rome—and by the derelict lands in Palestine, and Transjordania, once 'the land flowing with milk and honey'. The second, exploitation, is shown by the dust bowls of America—here was virgin soil, rich in natural humus; the utmost was extracted from the land, no living matter was returned, and consequently the life went out of the soil and it returned to dust. The results are being faced at last, and taken to heart with courageous enterprise and a stirring of national conscience. In New Zealand deep anxiety is expressed, because of the exploitation of land by the use of chemical fertilizers and of widespread deforestation. In

Introductory Note

Australia great tracts are suffering from drought, soil erosion, diminishing fertility from the same causes. From East Africa come accounts of virgin land exploited, doped with chemicals, till it becomes useless, then left derelict for a repetition of the same procedure a little farther on.

If we turn from the large to the small, we find the majority of small-holders and gardeners are up against the same difficulties. They cannot get natural humus, i.e. farmyard manure. They try chemicals—manure from a bag. It has all the right chemical ingredients, but no life, no inherent power of growth; has anyone ever heard of a mineral growing? The result after a few seasons is a steady decrease of fertility and increase of pests and disease. Mercifully, the compost heap is now being recognized as 'the heart of the garden'. This is a change of attitude of the past three years, and one which will surely save the situation, if the practice of using this compost becomes universal. In the midst of this world-wide sickness of soil, there are areas of fertility, and some in most unpromising natural conditions.

Primarily there are the Hunzas of Northern India; their valley home is an oasis of fertility, thanks to superhuman works whose origin must be in the far-off ages. Rock terraces hold the soil on arid hill sides; a system of irrigation, and —most important—the systematic and traditional making and use of compost have produced a race of human beings, healthy, happy and wise. Then in China, amid poverty and difficulty, the use and detailed care of the compost heaps form a definite part of community life. This has enabled the Chinese to extract the utmost from the same soil for thousands of years and still keep it alive and fertile.

At last there is a dawning realization throughout the civilized world of the importance, the urgency of this problem of soil fertility. Today, a growing body of people understand that the soil is a living thing and must be

Introductory Note

rightly fed. It is such common sense! All we eat comes from the soil, and derives its feeding qualities from the life in the soil. Meat, butter, milk, represent the vitality of the plants eaten by domestic animals. Vegetables and fruit give their vitality directly to us, but if they grow in unnourished soil, devitalized soil, they have no vitality to give.

The slow process of an almost universal malnutrition has started; it goes from soil to vegetation, from vegetation to human being. The result is a vast increase in malnutrition diseases—cancer amongst them; an increase in spite of modern amenities and the development of scientific knowledge, but knowledge that appears to be directed towards cure rather than prevention. The increase of bad health is not confined to man, it is shared by domestic animals, and by the vegetable kingdom. Every year brings the tale of new pests, new diseases, and new remedies—and insecticides. There *must* be a common cause for the universal symptoms, and the common cause of all that is—*is the soil*. If the soil is ill, all living beings suffer. The remedy must start there. Already proofs are available to show how a vast improvement in health has been brought about by feeding of the soil with organic composts, instead of doping the plants with synthetic manures. Evidence as to this has reached me again and again from Q.R. Compost users; and on a wider scale, the experience of schools and the well-known experiment of the Peckham Health Centre uphold the statement.

It is after all just common sense; common sense has been called 'heavenly wisdom', and a lack of it may lead to a world-wide tragedy, if steps are not taken to save the life of the soil.

I believe it is the force of public opinion that will tip the scales. There is much to overcome; vast vested interests; refusal to face facts; indifference and ignorance of urban

Introductory Note

populations; laziness and conservatism amongst the country folk; and the tentacles of a hundred years of synthetic manures.

An agricultural expert, who came to see the Q.R. Compost, and who was both friendly and appreciative, said to me, as he left: 'You know, Miss Bruce, we agricultural experts have *all* been grounded and brought up on the principles of chemical fertilizers and you can't expect us to change quickly'. That is true; but the change *is* coming and the increase of practical experience and personal knowledge will help to bring it about.

In 1939 I was discussing the title of a prospective book with the owner of a well-known nursery garden. I suggested 'Compost'. He just said: 'No, nine people out of ten wouldn't know what you meant'. He was right! Shortly afterwards. I was speaking at a garden fête on 'Compost'. An amateur gardener was asked why had he not attended as the talk was about gardens. His reply was: 'Gardens! I thought it was "*jam-making*"!'

Now the word is a commonplace; the value of compost is generally acknowledged in print and by authority, and, what is more, there is a widespread vocal revolt against 'manure from a bag'. The growing interest taken by doctors, hospitals, health centres, schools and other communities, shows how the wind blows, and the recent debate in the House of Lords, on 2nd February 1944, is a good omen for the future.

There are many methods of making compost and I believe there is room for all of them. There are millions of different gardens and different circumstances: if one method proves unsuitable, another may fit perfectly.

The three best-known biological systems, as quoted by F. H. Billington in *Compost*,[1] are:

[1] *Compost*, by F. H. Billington and Ben Easey (Faber and Faber).

Introductory Note

1. The Bio-dynamic method (Rudolf Steiner).
2. The Indore method (Sir Albert Howard).
3. Thd Quick-Return compost method.

All three produce good compost.

The Bio-dynamic system is interesting but complicated, and out of reach of the great majority, by reason of its restrictions.

For large farms and estates where there is plenty of livestock and labour, the Indore method is paramount. It is backed by the great knowledge and experience of its founder, Sir Albert Howard. But it presents difficulties to the small gardener without labour for turning or livestock for manure.

This book tells the story of the 'Quick-Return Compost' (Q.R.) for short!

It possesses three main advantages:

1. The compost needs no turning.

2. The vegetable matter disintegrates in an amazingly short time. Even after ten years' experience, I get the thrill of a miracle, every time I open a new heap, four to eight weeks after treatment, and find brown soil, rich in humus, instead of green vegetable matter.

3. The 'herbal activator' can be made by anyone who can find the right plants. The formula is given in full detail.

If it is impossible to find the herbs, the activator can be bought (see p. 83).

This is *not* primarily a money-making concern. It was launched in the hope of helping to give back life to the soil, and thus eventually of abolishing disease in plant, animal, and man.

This is a hope which can only be successfully fulfilled by the co-operation and personal effort of all who hold in trust a portion—however small—of God's earth.

CHAPTER 2

The Story of Quick-Return Compost

It all started with the garden, a derelict garden, but with beautiful bones. There were a few grand old trees, a lovely curved wall, and the rest was a wilderness, except for one strip which was planted with sickly cabbages. It stood on the very top of the Cotswold Hills. The soil was shallow and stony, thin, friable, and very, very hungry. The place had been a neglected farm. There was a yard full of ancient manure, and as long as that lasted, the garden did well. Then the manure gave out and I could get no more. In one season everything went back and I was in despair; I did not know what to do. I instinctively disliked the idea of chemical fertilizers, though at that time I knew nothing about 'compost' (this was nearly twenty years ago).[1] Then a friend told me about Dr. Steiner's method and the Anthroposophical Agricultural Foundation, as their English branch was called. My friend knew of it by hearsay, but it sounded so interesting that I got into touch, joined the Association, and acquired my first experience of compost and compost-making.

I learnt much of intense interest, accepted some of their theories, rejected more, queried the rest. I had some delightful experiences, pre-eminently a visit to Holland to see the Bio-dynamic Farms and the work of Dr. E. Pfeiffer. I learnt to appreciate the quality of compost and the effect it had on the land.

[1] i.e., about 1925.

The Story of Quick-Return Compost

But, as time went on, I realized that the need for compost was both world-wide and urgent, and I saw that it was the millions of smallholders, allotment-holders and gardeners who needed it most, for they were quite unable to get farmyard manure.

The Steiner Method seemed to me to be too complicated to have a universal appeal. The literature was too obscure. The process of making the 'preparations' used as activators is laborious, and though they can be purchased they are only available from a few sources, and at the time of which I am speaking could only be obtained by a member of the Bio-Dynamic Association.

There are of course many who use the preparations and rejoice in the methods. It always remains a question of individual appeal. I look back with real gratitude and much pleasure to their kindly friendship and all I learnt; but we gradually grew apart, and finally came to a parting of the ways, and I withdrew from the Association.

I was, of course, bound by my pledge of secrecy as regards the making of the special 'preparations', but I was convinced that there must be some *simple* way of reaching the same end, and making good compost, moreover a way which could and should be given to *all*. I told them of this belief, and that I should do my best to find some other method, and, when found, developed and proved, would publish it, and bring it to as many people as I could reach; and further, that as there was, and never had been, any secret about the identity of the wild flowers used in the Steiner method, I felt free to use the same herbs in my experiments.

There was a slight demur, but when I drew the Association's attention to the fact that, after all, it was *not* Dr. Steiner who had given either dandelions or nettles to the world, they could only laugh, acquiesce, and we parted

The Story of Quick-Return Compost

the best of friends, mutually wishing each other 'good luck'.

My boats were burnt; I can confess now that I felt very lost, completely blank, only believing intensely that an idea would come to my help—and come it did. I woke up one morning with the key to the problem in my mind and the words ringing in my head: 'The Divinity within the flower is sufficient of Itself'.

With the words came the understanding of what they meant: the life, the vitality within the herbs, in the sap. From previous experience I knew it had to be used in homœopathic quantities, according to the homœopathic creed of 'the power of the infinitely little'.

I started experiments that very day, extracted the juices from the living plants—dandelion, nettle, chamomile, yarrow, valerian, and made an infusion of oak bark.

The difficulty was to ascertain the right strength. I was no scientist; the only way was by practical experiment; and comparative tests. I filled a number of glass jam jars with lawn mowings, chopped-up weeds, nettles, and general vegetable matter. I treated them with the solutions in the following strengths:

1 in 10: 1 in 30: 1 in 60: 1 in 100—and then, urged by an impulse—1 in 10,000. There were two controls.

The jars were carefully labelled, then mixed and placed with the label towards the wall. Within five days the contents of one of the jars had gone ahead, and was changing colour rapidly. After ten days I invited a soil expert to come and see the progress of the experiment, and place the jars according to their merit. When he had made his choice, we turned them, label forward, and they read:

First, 1 in 10,000: Second, 1 in 100: Third, 1 in 60, and so on down to the controls which were still green, much as they had started. In fifteen days it was obvious that the

The Story of Quick-Return Compost

1 in 10,000 was far the best, in fact, almost broken down to compost.

What the jars showed was proved in the test 'heaps'. I took two numbers only—1 in 10,000 and 1 in 60. Again, the 1 in 10,000 was ripe and ready for use long before the 1 in 60.

I made one other, very crucial test. If this simple method was to be published, I must be certain that the compost was as good as the Anthroposophical one. So I made two identical heaps; treated one with the Steiner preparations, the other with the solutions. Both heaps were, or seemed, very good. I have not much faith in chemical analysis as a criterion of true compost value, but I sent a sample of each to a well-known soil analyst, and the returns were practically identical, with the comment: 'Of equal manurial value'. I thought *that* was good enough.

From then, it was the heaps that taught me the most valuable lessons. I had realized that heat was a vital part of the breaking-down process, and that the conservation of this heat was of utmost importance. To this end, wooden bins were made; they had no bottom, but stood directly on the soil. The cheapest form of timber in those pre-war days was old railway sleepers, 9 ft. long and 9 in. by 4 in. thick. They were everlasting, solid. Three half sleepers made each side, and three sleepers, one on top of the other, gave length and height. They could easily be sub-divided into any desirable width. The bins were built against a stone wall. The irregularity of their edges admitted air, and a roof of stretched hop-sacks kept out the rain.

The heap taught me how essential it was to keep an extra piece of sacking on the top layer *all the time it was being built*. One day a large corner of this covering was blown back, and that corner was stone cold while the covered portion remained hot and happy. I learnt the lesson: its importance cannot be overlooked.

The Story of Quick-Return Compost

It was the heap that taught me that if a large quantity of any one material is piled together, it takes a long time to break down, and in the case of lawn-mowings it packs together into a slimy mush. Hence the advice to make *no* layers thicker than four inches—and if possible to follow a layer of tough stuff with one of soft, juicy weeds, or cut grass—the one helps the other. I learnt too the importance of keeping the layers flat, by light pressure, so as to prevent crossing stems forming large pockets of air, and to ensure that the sides were packed up to the level of the centre. Heavy pressure is bad, but light treading or packing with a spade is beneficial, and I learnt that, in the bin, with level packing and the control of the natural heat of decomposition, the breaking-down process was *even* throughout the heap, right up to the sides. I learnt moreover that by the injection of the solution (the activator) the need of turning was eliminated, and the speed of decomposition increased, so much so, that a spring heap became soil, rich black compost, in from four to six weeks! A summer heap took from six to eight weeks; an autumn one from eight to twelve weeks, but a winter heap remained asleep, unchanged, and unchanging, till the surge of spring re-awakened the life in the earth. The quick ripening of the compost meant a great increase in the amount available for the garden, and the garden soon responded. The soil became richer, blacker, plants more vigorous, diseases vanished, the colour of flowers deepened and the flavour of vegetables improved. Many people visited the garden, tried the system, and were delighted with the results.

There were scoffers, of course, especially of the scientific, chemical-analyst mind. I came up against this type twice in quick succession: one was a science master in a boys' college, who openly scoffed at the idea of the homœopathic dose of 1 in 10,000 having the slightest effect as an activator.

The Story of Quick-Return Compost

The other was the agricultural expert of a Land Settlement Scheme which was started to provide allotments, equipment and advice for certain depressed areas—a grand bit of social work. One of the heads of the association had heard of the Q.R. method and came to see for himself. (Incidentally the association was spending thousands a year on artificial fertilizers.) He was delighted with all he saw, and departed with leaflets and samples of the compost to show the agricultural expert; naturally nothing could be done without expert sanction.

In a few days the agricultural expert's report was sent me, with deep regret and a request to answer it. The expert turned the system down utterly and completely. He said: (1) plants required certain carbo-hydrates which were not present in the solutions; (2) that if the method were adopted it would result in (*a*) very slow disintegration; (*b*) a compost of no manurial value whatever!

I answered the letter, pointed out that modern science recognized and utilized the forces of radiations, vibrations and emanations, *all* of which were beyond the power of detection by chemical analysis. It seemed a pity that agriculture—a science of 'life'—should deny the possibility of achievement along such paths. As to his two authoritative assertions, BOTH were disproved by practical experience.

(*a*) No one could call an average of two months 'slow' disintegration.

(*b*) My own flower garden had had nothing but vegetable compost for four years and the quality of its produce, the health of the plants, and the colour of the flowers were well known over a wide area.

I received a short, non-committal reply, and the matter dropped.

By then I was longing for some outside proof, some chance happening that would prove the value of the solu-

The Story of Quick-Return Compost

tions beyond all doubt; and my wish was to be granted in a two-fold manner.

I left home for a three weeks' holiday. Before leaving, I completed an experiment which I feared would prove a failure. I had a heap mainly of lawn-mowings, of which there was a surplus; they were put into a heap with about 25 per cent of dry leaves and soil, and *not* trodden down, as lawn-mowings make a poultice if they are pressed together. It had taken three weeks to build; I opened it before treating it, out of curiosity, and it smelt bad! I closed it, put in the solutions, left it, fearing the worst, and put it out of my mind.

During my visits, I went to a compost enthusiast, who took me straight out to see a new heap. It had been treated three weeks before (the month was August). It was not quite ripe, but it was getting friable, and it smelt very sweet.

'Now,' said my hostess. 'Come and open *this* heap. It was treated early in June, and it ought to be completely ready.'

I opened it. It smelt to heaven of decomposing cabbages! Awful! It looked slimy, green and yellow. The words burst out, 'This heap has *not* been treated.'

'But it *has*,' said my friend, aghast at both sight and smell. She called her gardener. 'Turner, YOU treated this heap, didn't you?'

In his slow Sussex voice he answered: 'No, marm, not that 'eap I didn't. I never touched that 'eap,' and on further enquiry it was proved that the heap had *not* been treated.

There was my first outside chance proof. The second was given on my return home! I went straight to the grass heap, left three weeks before as a slimy green mass. I plunged my hands into sweet friable compost, as good as anything I had ever seen.

It was the complete answer. From that day my confidence in the solutions has never wavered.

At this time, the solutions were seven in number, as honey

The Story of Quick-Return Compost

had been added at the same strength, 1 in 10,000. It is a powerful activator. The seven were kept in separate bottles, and inserted separately—a somewhat clumsy method.

Farmers were beginning to show interest and I realized that some simplification was necessary. I tried putting all the solutions together in one bottle. It proved absolutely successful, except that the honey was too lively and acted as a ferment. It had to be kept apart, till the final dilution for treating the heap; but the seven bottles were reduced to two, and the inoculation of the heap was accordingly simplified.

This led to a wider expansion and greater public interest. In 1938 Mr. L. F. Easterbrook, the agricultural correspondent of the *News-Chronicle* and an enthusiastic Q.R. compost maker of some years' standing, wrote an article describing the method and its results, with warm appreciation: hundreds of applications for further details poured in.

I then began to wonder *what* the power was that speeded up disintegration and produced such good results. I knew I had been working blindly, and that further knowledge was essential, if the method were to be really established. A book on herbs (*Nature's Remedies*) came into my hands by chance and gave me the clue. I found that the plants used in the solutions held between them *the chief elements needed by plant life*, and it dawned on me that these elements were in *living plant form*, and would therefore be of greater value than the same elements given in static mineral form by chemical fertilizers. (Can a mineral *grow*?) The list included iron, lime, soda, potash, phosphorus, sulphur, ammonia and carbonic acid. Further research at the library of the British Museum confirmed, and added nitrates to the list of plant constituents.

A discussion with an expert herbalist revealed some interesting facts. For instance: very few plants have been

The Story of Quick-Return Compost

analysed. The constituents of plants vary each year, not in kind, but in relationship to each other, according to the weather variations within the seasons. One year one constituent will predominate, the next year it may be another. I wondered might this not be a wonderful provision of Nature? The surplus, or lack of rain, sun or wind, in a given season, would have a definite effect on the soil; maybe cause a lack of some essential element. Therefore, Nature gives a little extra of this element to the plants, and as they disintegrate and return to feed the soil, they add an extra quota of the missing element, and so help to maintain its normal balance. If this were so, the practice of making a fresh vintage for the solutions every autumn would be wise, as it would keep the compost heap closely adjusted to the need of the soil for the coming season.

This line of thought prompted new experiments, to see if a successful activator (solution) could be made by using any two or three of the herbs that supplied, between them, all the chief elements.

I found that yarrow and nettle made a perfect combination.

Yarrow has: iron, lime, soda, potash, phosphorus, sulphur and nitrates.

Stinging Nettle has: ammonia, carbonic acid, formic acid and iron.

The heaps treated with these two solutions, plus honey, gave very good results, so good that I was tempted to scrap the full formulae, and use only these two: then came a further and unexpected development.

I had long realized that the activator worked by radiation. By no other means could the injection of a solution of the strength of 1 in 10,000 (approximately one drop to one pint) affect a ton of solid material. The process of injection is as follows:

The Story of Quick-Return Compost

When the heap is finished holes are made with a crowbar. These holes are from twelve to twenty-four inches apart, and reach to within six inches of the bottom. Three ounces of the diluted solution are poured into each hole, which is then filled with dry soil.[1]

The radiations start from these focal points, travel upwards and outwards and affect the whole heap. A London doctor, a pioneer in radio-therapy, visited me at this time, and was deeply interested in the heaps and the use of the solutions. He asked how they worked? I replied, 'By radiation; their vitality streams through the heap, conveying their living elements to every part of it, stimulating, vitalizing, energizing the whole pile, and all that is in it. I believe this vitality goes on into the garden, and into the plants that grow in it.' Then I added that vegetables should *not* be judged by size, but by their vitality, and there ought to be a 'vitality measuring' instrument for judging at every show! He laughed and said: 'I would like to test the solutions on my instrument, from the point of view of human health.'

He took a bottle of each of the pure essences, and wrote later that he found them to be the most *powerful factors for the destruction of human diseases*, and further, that each one affected a different disease, or group of diseases; and, please note, he was using their *radiations* only.

The outcome of the visit was twofold: First, I undertook to supply him with the essences, and have done so, in one form or another, ever since. Second: I reconsidered my decision to use only nettle and yarrow. They are the two *essential* herbs, but obviously, herbs possess some personal attributes as well as the elements they largely share. (They have been used in medicine from the beginning of time.) If these gifts are potent as regards human welfare, was it

[1] See page 38 for alternative horizontal method of application.

The Story of Quick-Return Compost

not possible that they might also be a safeguard against plant ailments?

It would be difficult to prove, and require far more knowledge than I possessed; but, with the possibility in mind, the full formulae could not be discarded.

Thus step by step the method has evolved, and last year, 1943–4, in its tenth year of existence, came what I believe to be the greatest step of all.

For two years I had been sending the herbs to the radiotherapist in the form of herbal powders. It had solved some technical difficulties and been very successful.

It struck me that if one could use the dry powder as an activator, it would simplify everything. There were difficulties to overcome; it took nearly a year to experiment, test, and get full and reliable results. But success came, and success beyond all expectation.

The activator now goes out in the form of a herbal powder, which is made of the seven ingredients *including the honey*. One grain weight (approximately enough to cover a sixpence or American cent) is dropped into one pint of water, shaken, and allowed to stand for twenty-four hours. This is injected into the heap in the normal way. It produces first-rate compost, just as good, if not better, than the original essences. It has been practically tested by several Q.R. enthusiasts, and received a cordial welcome. I believe it marks the greatest step forward so far in the history of Q.R. compost, and it entirely fulfils the directions of the words that rang in my head at the beginning.

The simple mixture of the plants and honey (which *is* an essence from the flowers) provides a simple agent for quickly turning vegetable waste into compost.

'The Divinity within the flower was and is sufficient in Itself.'

CHAPTER 3

The How and the Why of the Heap

How does a compost heap disintegrate? If you know the answer to this question, you will be a hundred per cent successful in the making of one.

The compost heap is one great co-operative workshop of living entities. Heat, the natural heat of disintegration, plays an important part. It comes from the quick breaking down of living tissues, leaves, stems and flowers; intense heat for a few days; then, with the release of the plant juices, it tempers to a moist pleasant warmth, ideal for the life and action of countless millions of microscopic soil workers. I repeat, *countless millions*, in the space of one teaspoon; bacteria, fungi, microbes (microflora), each one working at the further transformation of the vegetable matter, dying themselves, adding their minute beings to the sum total of the humus in the heap. They are supplemented by larger life, maggots, insects, and above all, worms, each with its own individual task; all working to turn the vegetable matter into food for new plant life. These beings need *air*. They must breathe; therefore, both aeration and the retention of heat are essentials for a successful heap.

THE BIN

To achieve this, we use a simple wooden bin: a box with four sides and no bottom. It stands directly on the soil.[1]

[1] Do *not* build on a concrete foundation and do *not* use creosote as a preservative—use Stockholm tar or old engine oil.

The How and the Why of the Heap

Why wood? Because it is warm, alive, generally obtainable, and easily erected. But there *are* substitutes, and in these days we may have to use them:

1. Oak staves of old barrels.
2. Brick walls, with spaces for aeration, say five a side.
3. Turf placed grass downward, and freed of squitch.
4. Bales of straw built round the heap.
5. If no protection is available, build the heap like a hay stack with straight firm sides. The inside will become compost. The outside six inches will act as protection; it will not decompose, but you can use it in the next heap.[1]

SIZE OF THE BIN

Suit it to the size of your garden. Aim at filling it within two months—the quicker the better; the fresh material shrinks tremendously, and a bin holds far more than you would think. Everyone is apt to start too big! It is far better to have two smaller bins than one large one, though you can always sub-divide a large one with light movable boards. A good general size is:

    For a small garden   18 in. × 18 in. × 2 ft. high
    For a medium garden 3 ft. × 4 ft. × 3 ft. high
    For a large garden   6 ft. × 6 ft. × 3 ft. high

Site of the bin—any aspect except north.

PROTECTION AGAINST RAIN

This is important because:

1. Heavy rain will douse the heat.
2. A sodden and confined heap cannot breathe. It is the aerobic (i.e., air-breathing) microbes that produce compost; the anaerobic microbes exist without air, and the result of their activities is putrefaction. Therefore, it is important to have adequate shelter to ensure both heat and air. Place

[1] *Never* use walls of corrugated iron or any other metal.

The How and the Why of the Heap

a sheet of corrugated iron at a slant, so that air can pass under and rain run off it: or, as an alternative, make a shelter of stretched canvas or strong sacking. Rubber is not advisable, as it is an insulator.

THE FOUNDATION

Good drainage is essential. If your soil is light, place the bin directly on it. If it is heavy, dig down about six inches and fill the space with rubble and a cover of soil on top. *Why?* Because the heap produces a lot of moisture, especially when plants are succulent. This must be able to disperse, or it would saturate the compost and exclude the air.

CHARCOAL

It is advisable, though not essential, to scatter a few handfuls of charcoal on the floor of the bin. *Why?* Because charcoal absorbs unpleasant gases, and remains itself unchanged. For this reason, it is given in the form of charcoal biscuits to relieve indigestion. It is also used in filters, and, in increasing quantities, in gardens, especially as drainage for pots and seed boxes. It is easy to make. Build a small bonfire, with brash wood (old pea sticks) and when it is red hot, pour some water on it—you will get charcoal.

BUILDING THE HEAP (MATERIALS)

Use any vegetable matter. Weeds, clearings of beds and borders, lawn mowings, cabbage leaves, vegetable peelings, tea leaves, coffee grounds, straw, old hay: animal manure, if you can get it. *Don't* use meat refuse, skin, bones, fat, or cooked stuff. *Why?* Because if kitchen refuse, other than vegetables, is admitted to the heap, you will get greasy water, greasy remains, in short—swill. Such grease makes a scum and keeps out the air, and that will lead to putrefaction. Also a daily libation of this refuse will over-balance your

The How and the Why of the Heap

heap, and the result, again, will be putrefaction, smell and flies. In a large *farm* heap with manure, kitchen refuse *might* be risked, but I strongly advise against it.[1]

People may say: 'But animals that die go back to soil!' Yes, of course they do, but most wild animals do not die a natural death. They are the prey of others, right down the scale. Hunt for the body of a dead beastie, hunt through an acre of woodland—you may find one, possibly two, but I doubt it, and *we* are dealing with an area of a few square feet!

Following this line of thought, I have heard of people getting offal and remains from the butcher's refuse, as a weekly offering to the compost heap; but again, to do so would be to over-balance the heap, and go beyond Nature's own scheme. One or two odd mice or birds buried in a heap will disappear, and be absorbed by the mass of vegetable matter and the work of the micro-organisms, but for a garden heap, I counsel no weekly offerings of flesh and *no* metal. It is not just a rubbish heap!

WEEDS

Use *all* weeds, even seeding and rampant ones. Place seeding weeds in the centre where the heat will destroy their power of germination. Have no fear of rampant weeds. The more they ramp, the more vitality they have to give to the heap. Better not put them near the top; in late autumn, they may grow to the light, but they will not root, and can easily be pulled out and used on the next heap. I am thinking of bind-weed, a bad ramper, but it disappears entirely in the heat of a heap. The only plants to avoid are heavy tough evergreens, i.e. old ivy leaves, old privet, and yew.

---

[1] I suggest that this is a matter of individual experimentation—I use all kitchen refuse in my own compost heap, and have never found it gave any trouble, but in our household it contains very little meat. Editor.

The How and the Why of the Heap

Use the green stuff as fresh as possible. The fresher it is the more vitality it holds. If you can't use it at once, throw a sack over it, to prevent sun and wind drying it up. If it seems shrivelled, spray it before adding it to the heap; cut your long stems into short lengths, six to twelve inches. *Why?* It releases the juices and the short bits pack better. Use a sharp heavy spade for this job, which is soon done. Incidentally, if a stem is too tough to be severed with a spade, it is too tough for the heap. Burn it.

BUILDING THE HEAP

Build in layers four inches thick. Alternate layers of tough stuff with soft green weeds or grass, the one will help the other. If you have animal manure, or poultry manure, put a two-inch layer or less, in every foot. If you have none, throw in a scattering of soil. Introduce three dustings of lime. I repeat *dustings only*, at twelve, twenty-four, thirty-six inches.

Keep the heap level. At first it will tend to build up in the centre and sink at the sides but once heating sets in, the reverse tendency takes place; a *light* treading or packing with a spade will correct this. It will also break down crossing stems, which make air pockets.

Always keep some sacking on the last layer. This is very important. *Why?* Because sun and wind dry up and shrivel the exposed area, and heat, moisture and vitality escape from the heap. This heat can be intense, it reaches 160 to 180 degrees for a short time, then dies down; it rises again when fresh material is added. To maintain a steady heat make new additions as often as possible. Decomposition is quicker, and the intense heat destroys weed seeds and disease.

The heap will shrink tremendously as you build it. As long as there is heat in it, you can go on adding fresh material.

The How and the Why of the Heap

When it is full and firm, cover it with four inches of soil, let it settle for two or three days, then treat it with the 'activator'. Unless horizontal method has been used (see below).

THE TREATMENT

The activator comes in the form of a herbal powder (formulae, p. 79). Drop one grain (a pinch, as much as will cover a sixpence or an American cent) into a pint bottle of rain-water. Shake it well; let it stand for twenty-four hours. Shake again before using it. It will keep in solution for about a month or three weeks. If it smells sweet it is all right.

INOCULATION

Vertical Method (Best for home made powder)

Make holes with a crowbar from approximately twelve to twenty-four inches apart, and to within three to six inches of the bottom of the heap; pour three ounces of the liquid into each hole. Fill them up with dry soil, and ram it down to prevent air pockets. Cover the heap with a sack, and forget it for a month.

Horizontal Method (Best for purchased powder see p. 81).

Activate the heap as you build it, by lightly sprinkling each 4 to 6 in. layers with the Q.R. Solution. Allow about $1\frac{1}{2}$ oz. per layer, approximately 1 pint per heap. The amount used is very elastic, a little more or less makes no matter but—avoid *saturation*. Cover the sprinkled layer at once with fresh material, i.e. do not sprinkle the last layer of the day till you are ready to continue building. The heap is breaking down while you are working at it, the activation will be thorough and even.

Method of Application. Use a pint bottle with a sprinkler cork, such as housewives use for damping clothes, or cut a wedge in a cork, or slightly loosen the cork in a screw-top

The How and the Why of the Heap

bottle. For a farm heap, mix a dessertspoonful of powder in a gallon watering-can with a fine rose—stir it *well*.

RESULT

When you open it, burrow into it with a trowel. If it smells sweet (and it has a lovely smell) it is all right: dig further, breaking it up as you go. If rightly built it will be very rich dark soil.

Remember, it is impossible to give a *definite date* for the ripening of any heap. There are so many differing factors: season, weather, building materials—one can only give an average and approximate time. Roughly speaking:

A spring and early summer heap takes four to six weeks. A summer heap six to eight weeks. An autumn heap eight to twelve weeks.

A winter heap moves very little if at all. The earth sleeps in winter and this seems to affect both growth and decay. You may make a wonderful heap of winter weeds, between 21 December and March: it develops no heat, it just remains as you put it in. *But* when you get some fresh spring growth, or, best of all, the first lawn-mowings, remove the top half of your winter collection, introduce a four-inch layer of the living green, and build up the heap in alternate layers of winter weeds and fresh growth. *That* heap will decompose in about a month, and you will get the advantage of increased bulk with the help of the winter collection.

If, when you open your heaps, you find they are not entirely soil, there is usually a reason, and always a remedy.

1. You may find a sodden corner, or possibly a sodden layer. *Why?* Probably rain has seeped in, or it may be after a wet spell your plants were full of moisture which could not drain away. *Remedy.* Let it remain in the air for a few hours, it will soon disintegrate.

2. It may be that some stems or tough grass have not

The How and the Why of the Heap

broken down. *Why?* Probably they were too wiry, too dried up; dry old grass is difficult. *Remedy.* Fork them into a loose pile in the open, and pour some compost or manure water (see p. 54) over them; in a couple of days they will be fit to put on the garden.

STORING RIPE COMPOST

When the compost is ripe and you need the bin, you can stack it in a compact heap, with steeply sloped sides, covered with soil, so that rain will run off. It will go on ripening and come to no harm.

AUTUMN LEAVES

Do not use fallen leaves in a mixed heap. A few odd ones don't matter, but a thick layer makes an impenetrable barrier and holds up the heap. *Why?*

1. Because their flat surfaces press tight together and exclude air.

2. Because they have already lost much of their vitality. They are half dead, or they would not have fallen, therefore they decompose more slowly than living green matter and they slow down the decomposition of the whole heap. They make first-rate humus, but it is better to make a separate leaf-stack, in the open. A surround of wire netting keeps it tidy. An occasional scattering of soil is all to the good—do not tread it down. Leave it for six to nine months. Then turn it out and treat it. In two months it will turn to a very rich black mould, like 100-years'-old leaf mould. Once you start a rotation, you need never be without it. It carries not only its own leafy smell, but the added aroma and richness of the Q.R. compost.

MANURE

Some compost makers believe that animal manure is *essential* to making a good compost. I do not agree. I have

The How and the Why of the Heap

tested entirely vegetable compost, against compost made with manure, and found no difference in its effect.

After all, cow manure is just plants, composted by the cow. She is the best compost-making machine in the world! She breaks up the herbage by the combined heat of her body and the incessant chewing of the cud. She withdraws the vitality of the plants into herself, for her own needs (and ours), bones, blood, flesh and milk. She has three stomachs to complete this work and should do it thoroughly. What she cannot assimilate, she returns to the earth as dung, i.e. composted plants. (Note that the smell of cow dung, and the smell of rotting lawn-mowings are almost identical.) But this cow-made compost is full of very strong animal digestive juices; if it is used in the garden *when it is fresh*, it burns plants, and introduces pests into the soil; but, if you wait two or three years, it turns to a beautiful black soil, like the best compost.

Now in the vegetable compost, we use the entire plant, with its vitality whole and unimpaired. We have learnt from the cow! We use both pressure and heat, but instead of the digestive juices we use the herbal activator containing the chief plant elements, in *living* plant form. Moreover we beat the cow at her own game, as regards time! Instead of having to wait for two or more years, the vegetable compost is ready for the garden in six weeks—or less!

If it were true that animal manure is essential to good compost it would be a tragic outlook for millions of gardeners and smallholders. Even before the war, I found that between 85 and 90 per cent of gardeners were unable to get farmyard manure for their holdings.

I arrived at these statistics by questioning the audience, at every meeting I addressed. I called for 'Hands-up' from all who could get farmyard manure for their gardens. The average was always the same, with one exception—a private

The How and the Why of the Heap

meeting for farmers' wives in an entirely dairy country.

The lack of farmyard manure was so universal, that I decided on a drastic test.

I excluded all manure from the compost heaps designed for the flower garden, using *only* vegetable compost for four years, and I found that soil, health, and beauty of growth and colour were not impaired in any way. I further know from reports that purely vegetable compost has had, and is having, splendid results in all parts of England; and it is the greatest comfort to garden lovers, to know that they need not be dependent on something they can't get!

Instead of liquid manure, they can soak a trowelful of ripe compost in a pail of water, dilute it to tea colour and use it as a feed for the plants that need it. The response is amazing.

At the same time, while not *essential*, manure *is* a help. It has wonderful heating and activating qualities and once it is ripe it is the finest natural humus that exists. A farmyard manure heap treated with Q.R. activator becomes ripe and friable in a few weeks (again we help the cow!). For those who cannot get it in bulk, it is possible to make a little go a very long way. Thus:

THE MANURE TUB

Sink a tub, barrel, or box, in the soil, to within six inches of its rim. Fill it to earth level with fresh cow dung. (Most farmers will allow you to collect a few pailfuls, from gateways and byres.) Treat the manure with three ounces of the diluted solution; cover it with a wooden lid to keep out the rain. It will be fit to use in about three weeks, and you use it for liquid manure. A trowelful in a gallon pail of water makes a strong brew. One pint of this to one gallon of water is the best strength for tomatoes or any plant needing food. Further, a bucketful of this, or the stronger liquid, can be poured into a ripe compost heap and will add to its richness.

The How and the Why of the Heap

A curious point about manure treated in this way is that though it loses its rank smell, it preserves its fresh appearance for years, and one barrelful will last a very long time.

Another method is to fill the sunk tub with *dry* cow-pats; treat them in the same way; cover them, and forget them for three months. When you go back to them they will have crumbled into the finest black soil, perfect for top dressing, but not for liquid manure.

This was a purely chance discovery. An order was misunderstood, and a tub, meant for fresh manure, was filled with these dry pats. Labour was scarce, time scarcer, so I left it, and treated the tub, just to see what would happen, and a miracle happened! Several experts who saw the results this summer pronounced it some of the best stuff they had ever handled and could not guess what its origin had been.

POULTRY AND RABBIT MANURE

With the war, there has been a tremendous increase in domestic poultry and rabbit keeping; consequently many appeals come from compost makers for advice in handling poultry and rabbit manure.

While very light layers of poultry manure can be used directly on the compost heap, we find the most satisfactory way is to make a separate small heap, like a miniature dung heap. We use the droppings, the litter, straw and hay. The dry straw is thoroughly wetted *before* building it into the heap. For this we use the strong manure or compost water. We build the heap up to two and a half feet, protect it from heavy rain, throw a spadeful of soil over it at intervals, and treat it with the solution. It breaks down in less than a month, and looks like farmyard manure. We put this on to the compost heap in two-inch layers. It makes good stuff. Rabbit manure could be treated in the same way, either in a separate heap, or with the poultry droppings.

The How and the Why of the Heap

Farm Heaps

It is obviously impossible to have bins all over the farm; therefore, farm heaps must be built in the open, and, as farm material is brought in by the car-load, instead of the barrowful, they must be of larger dimensions. A section eight feet long by six feet wide and six feet high is a useful size. One section can be completed before going on to the next; the sections can touch, and so make an ever-lengthening clamp. If the top is sharply ridged rain will not seep in. The procedure of building is the same as for the garden heap. Good drainage is necessary. Build in layers of four inches. If there is a mass of one material, break it by narrow layers of soil, or better still, manure. This should be available on the farm and can be used in two-inch layers throughout the heap. Material like old dry hay, tough grass, and above all, dry straw, should be saturated with treated manure water.

In Eire the Ministry of Agriculture advises soaking straw for twenty-four hours. A nursery gardener, who runs an 'intensive' garden with Q.R. compost, told me recently that he used a quantity of straw in his heaps, and soaked it overnight in a long bath filled with manure water. The results were first-rate.

If a farm is equipped with a urine tank, the tank itself can be treated. Soak some sand, or dry soil, in the diluted Q.R. solution, allowing one pint to each six cubic feet of tank space. Scatter the soaked sand over the surface. The sand will sink, and free the solution to do its work from the bottom. Straw soaked, or even sprayed, with this urine, would make valuable compost, and break down very quickly.

In an all straw heap, include if possible two-inch layers of fresh green nettles or bracken. The green gives vitality; nettles, wetted and bruised, will raise heat more quickly than

The How and the Why of the Heap

anything! Manure, if possible, otherwise soil in narrow layers, will steady the heap. Treat it; it will go to rich black mould, without turning, in from four to six months. It can then be spread with a shovel.

The method is very elastic and open to infinite variations. The three chief rules are:

1. Keep heat in.
2. Keep rain out.
3. Let the heap breathe.

ALTERNATIVE METHODS

During the intensive and rapid building of farm heaps some previous conclusions were emphasized, and some alternative methods proved to be successful.

Heat

I further confirmed that the very intense heat lasts for a short time only—at the most two to three days—then it dies down quickly, unless a fresh layer, however small, is added before it begins to decline. If this is done the heat passes straight up to the new addition and the intensity is maintained without a break. This is important, because if the heat drops below a certain point, the new layers have to restart it themselves, and it will take very much longer. This may be a counsel of perfection, but it explains the difference in time between the ripening of two seemingly identical heaps. It also emphasizes the point, so often urged, 'build as quickly as possible'.

Another suggestion for increasing the heat is to put weights on top of the protecting sacks. I discovered this by chance. I inadvertently left a heavy stone on top of a heap, and found next morning that the heat below the stone was greater than anywhere else, and that decomposition was further advanced. Now I systematically place either flat

The How and the Why of the Heap

stones, or short boards, weighted with a stone or a bit of metal, on top of the sacks, with quicker results.

Texture

I have always stressed the importance of mixing textures, by building in layers, but an alternative method, suggested by Mr. King of St. Leven's Hall, Westmorland, proved successful, and is simpler. The fresh green refuse is roughly chopped, and mixed all together *on the ground before* it is built into the heap. It is quicker, it packs better, it heats more quickly, and it eliminates the possibility of too thick a layer of any one texture. I offer it as a good alternative.

A HEAP IN THE OPEN

I experimented with the 'open' heap, and got very good results. Full success *lies in the method of building*. Mark out the space for your heap, 3 ft. to 4 ft. square. Build *each layer from the edges, to the centre*. Place the first forkfuls round the four outside edges, thus outlining the heap, the second lot just inside and overlapping the first rows, and so on till the final forkfuls of the first layer go on to the centre, and steady and firm the whole foundation. Build each layer on this pattern, and you will get straight firm sides like a well-built rick, and the heap will never slump into a loose, untidy pile. Build up to 3 ft., then diminish your layers into a dome-like top. Buttress the base of the heap with some soil. Protect the sides and top with grass clods (bull pates and the like), grass down, roots up. They will cling to the sides and eventually turn to first-rate loam. Keep a sack over the heap when you are building it. The experimental heap built on this pattern developed good heat and turned to compost in five weeks. It was good quality, and free of weeds.

The How and the Why of the Heap

Texture and Treading

There has been some discussion about the wisdom of treading the heap, and some users have disregarded the warning on p. 79 about too heavy treading. The whole question, to tread or not to tread, depends on the *texture* of the materials put in the heap. With a mixed heap, i.e. fibrous stems and soft material, light treading is essential. The stems crossing each other make large air pockets; the soft green disintegrates quickly and shrinks, enlarging those pockets, thus the heap becomes honeycombed with vacant spaces unless it is trodden down and close contact is assured. There is no fear of lack of oxygen from the close contact, because the quick disintegration and shrinkage of the soft green creates innumerable *small* air spaces which hold enough oxygen to maintain the air supply throughout the heap. The only thing that completely eliminates air is saturation by water. On the other hand you must *never tread a heap made entirely of soft material*, i.e. lawn mowings or farmyard manure, with no fibrous matter; such material would pack into a close poultice-like mush, and eliminate all air spaces.

Liquid Manure

I have had two interesting reports about the action of the Q.R. Activator on liquid manure.

1. A gardener had filled a barrel with bedroom slops including urine, hoping to use it on the garden. It developed an appalling stench. He soaked some sand in the diluted Q.R. liquid, scattered it over the surface of the barrel, in twenty-four hours the smell had gone; at the end of a week he used it on the garden, with first-rate results.

2. A well-known farmer wrote the following account of treated liquid manure. 'The pit holds a month's drainage from the cowsheds. I treated the pit when it was about three-

The How and the Why of the Heap

quarters full. There was a considerable amount of gas formed which was released on stirring the semi-solid matter at the bottom, and in three weeks' time all offensive smell had disappeared.' The contents were used on a field.

Straw

I might add a detailed account of the latest results in turning straw into humus. The heaps were built late last autumn. They were treated at the end of December, opened the 16th May, and were found to be rich, black and full of worms. The main method of building was as described on pages 46, 48, and 79, with this difference. The straw was in 6-inch layers. The dividing material was cattle-trodden soil, or manure, or netties. The saturation of the straw was by hand splashing from a pail. The activated manure water (see p. 54) was at the strength of:

Stock solution: 1 gal. activated cow manure to 20 galls. water.

For use: 1 pint to 1 gall. water, i.e. diluted to tea colour.

The stock solution was in a hand water barrow, the diluted liquid in two feeding bins 2 ft. in diameter. Thick bundles of straw were placed on top of the heaps as protection against rain.

OTHER MATERIALS

While the foregoing is about compost making by the more ordinary materials, there are some people who may have unusual, yet priceless raw matter, within easy reach —perhaps thousands of tons of possible compost—going to waste.

An interesting example is the story of a friend who lives near the New River, the chief water supply for London. Twice a year water-men clean the river of water weed, mud and the heavy growth of its banks. The water weed, green and crinkly, has untold vitality. It cannot be used on the

The How and the Why of the Heap

land for seven years, or it would start growing! It smells like pig manure; the river mud smells worse. The water-men pile it up in huge dumps and leave it. No one thought of using it, till my friend, a keen 'composter' and gardener, had the inspiration to try it.

The first heaps were made entirely of the water weed, in various stages:

1. The fresh weed as it came out of the river.
2. The slimy stuff, a week old, from the bank; and
3. A very small proportion of the dry seven-year-old rotted stuff.

With these ingredients, a layer of lime, and some layers of soil, several heaps were made, covered with earth and treated with the Q.R. 'solution'. In fourteen weeks, they had rotted to a friable dung, not good enough for top dressing, but good for putting into trenches to retain moisture for peas.

The next experiments were an even greater success.

The heaps were made with water weed, straw, and chipwood bedding from a large stable. The water weed wetted the straw, while the chipwood bedding, which had horse manure in it, made a dry, steadying layer.

Several such heaps were built and treated, and in three to four months had turned to a rich black compost, of first-rate quality. It produced one and a half tons of onions, and six cwt. of fine peas on one-half acre of poor land, and this in a very dry season, a universally bad one for peas.

The original heaps of treated water weed are now, after fifteen months, good black compost, described as 'like Lincolnshire silt'. The *untreated* water weed dumps *seven years old* are not compost, but described as 'a useful rather dirty muck'.

Thousands of tons of this first-class potential manure are wasted—and the land is hungry for it.

The How and the Why of the Heap

There is seaweed, especially bladder-wrack, and that broad ribbon-like bright tawny seaweed, often yards in length; build these into a mixed heap, place them between layers of fresh green material. They will break down quickly and are full of valuable living fertilizers.

In every country there must be waste products, tremendous growths, overwhelming weeds, which could be turned to compost, with imagination and a little care. Anything within the vegetable kingdom will turn to soil, with pressure, heat and aeration—and the earth needs all we can give her.

SUMMARY

The more one makes compost, the more one realizes how elastic it is, how many ways are right, how few are wrong, there is always something to learn, and it can be adapted to every condition.

I have had good reports from South Africa, Orange Free State, Transvaal, Rhodesia and Kenya, from Australia and New Zealand. They range from small gardens to vast farms; they show that if we resolutely follow Nature's basic laws of life, we can cooperate with her, and increase the fertility of the earth no matter *what* conditions we meet.

CHAPTER 4

The Compost and the Garden

The building of the heap may sound laborious! As a matter of fact, it is very simple, it soon fits into the routine of garden work, and always carries with it a sense of anticipation. Opening a ripe heap never loses its thrill of amazement! The change is so dramatic, the aroma so sweet and satisfying, the soil so clean and vital, one cannot keep one's fingers away from it. But I am not sure that the results of using the compost in the garden are not even more astounding.

It gives the soil whatever quality it lacks, no matter what it may be! It transforms heavy clay into friable mould; it gives thin soil substance, and hungry soil food. It suits every kind of plant. A clever Dutchman, an expert on soil, explained that point, in his slightly broken English: 'But—of course—do you not see? You are not giving to your plants *one dish*. You are offering them a restaurant and *they can choose for themselves*.'

It is easy to use. Keep it in the top four to six inches of the soil, and either fork or 'cultivate' it in. It *is* soil, and will amalgamate quickly with its surroundings. Use it at about six tons per acre; the equivalent is two and a half lb. per square yard. The quantity is approximate. You can't hurt the land by using too much. It will never make it sour, so you can be generous to hungry land and to hungry plants.

In intensive gardening, where crop follows crop, fork in a fresh dressing of compost, as you plant each successive one

The Compost and the Garden

(other than roots, of course). If you are short of it, give a little to each individual plant.

A good mixture for seed boxes and frames is

Loam, 5 parts; compost, 2 parts; sand, 1 part.

For seed drills: line with a sprinkle of sifted compost and sand.

Tomatoes in Pots

Start with the normal mixture, loam 5 parts, compost 2 parts. Plant low in the pot—the soil level about half-way up. When the roots appear on the surface put two inches of top dressing:

1st dressing: 4 parts loam; 3 parts compost.

2nd dressing: 3 parts loam; 4 parts compost.

3rd dressing: 2 parts loam; 5 parts compost.

When the first truss sets and is the size of a golf ball, use pure compost. When the fruit is ripening, give either manure or compost water, every ten days. When the roots grow out of the drainage holes, put compost on the stage. This method has proved most successful.

Special Uses

For peas and beans we mix a layer of compost into the top of the second spit to hold the moisture, and of course add the usual dressing to the top soil.

It makes a perfect mulch for soft fruit, greedy vegetables, and wall fruit.

For feeding orchard trees, we make up a strong brew of 'manure water' from the fresh manure tub (see p. 80) (or compost water) one gallon manure to twenty gallons water. We fill a wheel water-barrel with the strong mixture, dilute it to tea colour for use (roughly one pint to one gallon of water). We run a fork straight into the soil, at three-foot intervals, round the outside stretch of the branches and pour

The Compost and the Garden

one gallon of the liquid into the tiny holes. This is done in early spring.

The result is amazing. Old tired trees (the orchard is very old) have taken a new lease of life, with vigorous healthy foliage, and bear outstandingly good crops.

Before the war I had a staff of three men: a chauffeur-handyman, who also mowed the lawns, and two men in the garden. There is a kitchen garden of three-quarters of an acre and two glass houses, besides many fencing jobs on the farm and round the woods. The place is 150 acres, of which 50 are woodland. The farm was let for grazing, but the woods and fences are my responsibility. The fences are mostly dry Cotswold walls, so the men were busy.

My own garden work was entirely the flower garden—now a place of beauty. The whole of the enclosure is protected on the north side by the curving grey wall. It was always a vista of colour bordering a broad turf walk, curving in harmony with the wall; green lawns were dominated by a group of trees, including a copper beech, and a marvellous cedar; all this in a setting of woods and hills with a view down the Golden Valley giving both vista and space.

The sole care of this garden, plus the compost work in all its branches, filled most of every day. The kitchen garden, and fruit, was left entirely to the gardener. He made his own compost, at first grudgingly and without much care. He was 'set in his own ways' and his results were not first-rate, till one day, he confessed he had no compost fit for use—what was he to do? I told him to use one of my heaps, from the 'garden' yard, and left him to it. When I next saw him his face was wreathed in smiles. 'The best stuff I have ever handled, Miss!' After that, there was more care and better results.

One half of the kitchen garden had been literally carved

The Compost and the Garden

out of a field. The soil was originally two inches in depth, then came rock! We pick-axed it out, three feet deep—pure Cotswold stone—tons of it—and then had to fill up with whatever we could get. It came from old field dumps, from the woods, and the roadsides; a stupendous task and a weird mixture, but the compost pulled it together and today it is a good garden, though still a stony one.

In 1940 the whole staff joined up, and in their place I had one *very* old man; an odd job labourer, well over seventy, crippled in both hands, and lame. A gallant old fellow, who talked such broad 'Gloucestershire' that at first I couldn't understand him, he was a first-rate stone-waller (a Cotswold craft), good at straightforward digging, but beyond that not much of a gardener. The cry then, as now, was 'Dig for Victory', so all my efforts went to the kitchen garden, and the flowers had to look after themselves.

I had a lot to learn and I loved it. I made and used far more compost than the garden had been having and some of the results were startling.

STRAWBERRIES

I had been told that strawberries would not grow on the Cotswolds, certainly they had never succeeded hitherto. I divided some old plants, took a few runners, made a rich bed, heavily composted, rammed the friable earth till it was firm. (They like hard planting.) After the first year they were a wonderful sight: healthy, large plants roped with berries, immense and very sweet. Everyone coming to the garden would stop, and gasp, 'Oh, LOOK at the strawberries!'

This is an outstanding case, but looking back on the garden as a whole, the general appearance of the crops has been consistently good. I remember a delightful instance—before the war. A party of professional gardeners came to see the compost and its results. They started politely but frankly

The Compost and the Garden

sceptical; when they reached the kitchen garden there was a silence: we came into view of the asparagus bed (the month was September). A voice said: 'Coo! Look at they! I thought they was young larches!' We went on and there were a few murmured words of appreciation. When we had done the rounds, the leader, with a charming gesture, took off his hat and said, 'Well, Miss—I have learnt something. You often see one or other good healthy crop, but you very seldom find *all* the crops equally healthy. You've got that, here. There's something that's all *right*. Thank you.' He was an old man and it meant a lot. Incidentally, one of the party became a gardener to a nursing home on the Cotswolds and the Q.R. system is used as a matter of course—and success.

WEATHER RESISTANCE

This is important in England and the resistance to extremes of weather was proved in the severe winter of 1940. Frost, a partial thaw, then a fine rain that fell—and literally froze as it fell. Every blade of grass became a column of clear ice. Every twig was coated with inches of ice. Birds froze on the trees. Branches crashed every few minutes, borne down by sheer weight of ice, and in falling broke those beneath them. We woke to a world of clear ice, great beauty, and appalling devastation.

Visiting gardeners during the following summer told me they had lost 90 per cent of their brussels sprouts. The crop in this garden came through undamaged—and the vegetable garden faces north at an altitude of 750 feet.

Resistance against drought is also noteworthy. In this stony friable soil, we would welcome rain every week! But now the soil holds the moisture and the plants flourish even in a prolonged drought.

The Compost and the Garden

PEST RESISTANCE

One of the happiest results of this vegetable manure is increased resistance to pests and disease.

It is common sense, of course! If plants (like humans) are well fed and full of vitality, they withstand diseases. Disease or pests may show themselves, but they do *not get the upper hand.*

Even in the worst cabbage butterfly years, when caterpillars reduced plants to skeletons, only a few scattered holes showed amongst the compost-grown brassica. One theory is that the marauders eat a little, and are satisfied, while with devitalized plants, 'they go on eating, *seeking for something that is not there*', till they have destroyed the whole.

The 'Old Man' soon became a convert; his wife is a good gardener, and together they made compost for their own plot. Last year onion mildew was rampant in his village, and he told me that his two immediate neighbours lost the whole of their crops: his rows escaped without a touch of the disease. He added that he had only been able to compost half his row of beans, and—'Thee could'st tell to an inch, where us left un off.'

A correspondent from Nottingham wrote that ground pests had played havoc all round the countryside. His plot was in the centre of three ravaged gardens—yet it came through unscathed, and, what was more striking, a friend of his lost every plant in his garden *with the exception of some cabbage plants, raised on compost, and given to him by the writer.*

An equivalent to this happened here during a season when blue aphis was prevalent. After a careful survey, I found only 3 per cent of my brassica were attacked. Of these only three were my own plants, in each case weaklings with double

The Compost and the Garden

stems, that should not have been planted out. The rest were plants given to me from a non-composted garden.

So the evidence grows, and is repeated from many sources. Now—what about quality?

QUALITY

Size alone is nothing. The supreme test is taste, texture, and feeding qualities.

I recall a young visitor who blew in one morning to see the garden and the compost. He was an enthusiast—almost sang the praises of a certain shop that sold compost-fed vegetables. Never was there such food! It was a revelation! We discussed this and many matters till the morning flew, and the lunch bell rang. 'Come in and take pot-luck,' I said. He accepted, and we sat down. The vegetable was spring cabbage. He was talking eagerly. Suddenly he stopped short and realized what he was eating; he took a second mouthful, then a third, in complete silence. Then he laid down his fork, leant over the table, and said impressively, 'Miss Bruce, this has the— beat to a frazzle!'

It is a curious fact that while ordinary cabbage smells when cooking, a compost-fed cabbage does not smell at all. There is a delicious, subtle sweetness about all the vegetables and fruit—a curious reminder, in taste, of the aroma of the compost itself. It is noticed, and commented on, by all visitors (even those who know nothing about the garden). It is eagerly recognized by those who use compost themselves. It is also discernible in the honey from the garden hives.

A passing visitor from Birmingham told me this story. He had been a keen Q.R. compost maker for three years, and was convinced of the different taste, and better quality of his own home-grown vegetables but, fearing his enthusiasm, he wanted an outside and conclusive proof. So one morning, instead of getting the potatoes from the garden, he secretly

The Compost and the Garden

bought them from the greengrocer. At the midday meal he watched his wife. She ate a little, looked puzzled and dissatisfied, and finally said, 'What potatoes are these?' He answered grimly, but truthfully, 'They came from a different row.' After a time, when she showed obvious distaste, he remarked, 'You don't seem to be getting on with your potatoes, my dear. What is wrong?'

She burst out, 'Well, they are entirely different from the ones we have been having. They have no taste, they are exactly like the stuff we used to buy at the shop!' He had his proof. He confessed his trick, and both are more compost-minded than ever.

From a commercial point of view this aspect is very valuable. A smallholder specializing in compost-grown vegetables should, and does, command top prices.

One composter told me that a local greengrocer was so struck with his produce of his garden, that he offered him retail prices for all the vegetables he could supply.

A smallholder, running his garden on the French intensive system, wrote that he had solved the problem of the early market by the combination of 'Cloche and Compost'. He had got ten shillings a pound on open market for very early peas.

The comment of the tenant-farmer who now tills my land, is illuminating: 'There's something in it! Here's these fields, just the same land, let for twelve shillings an acre—and there's the garden growing crops as if it were land worth five pounds an acre!'

A chance seed of wheat was dropped by a bird in a corner of the garden. It grew magnificently, had forty stems like young bamboos, and yielded two ounces of seeds.

The farmer lives ten miles away and has to bring his team of workers by lorry. He has not the available local labour to make farm compost heaps; but after threshing, my 'Old

The Compost and the Garden

Man' made up several heaps of straw, cavings, weeds, and a little manure. They were treated, and the farmer tried them on a breadth of field, to be sown with oats. The stooks on this part of the field stood a foot higher than the rest, and the grain was infinitely heavier. The foreman was thrilled, and he and several of the hands begged for some compost, to put into their own gardens.

FEEDING QUALITY

This is the most important matter of all. If a plant is healthy, and growing up to its own perfection, it must have great vitality, and it is the vitality, the living force of the plant, that heightens its food value. The satisfying quality of the vegetables is noticed by all visitors, and is of value in these days of rationing. A vegetable cannot give what it has not got; what it has, it gets from the soil. It cannot reach its 'own perfection' in starved ground, still less in ground doped with chemicals.

Most artificials are soluble salts,[1] so strong that a warning is given to avoid touching the foliage, for fear of burning it. The salts dissolve on the damp soil, or are watered in. The unfortunate plants are bound to absorb them. A burning salt solution! What harm can it not do? It may act as a stimulant, but to feed on a stimulant eventually ends in weakened constitution, disease and disaster.

---

[1] Lime is not an artificial fertilizer. It is a natural mineral. It does not dissolve and feed plants directly. It sweetens the soil and helps to release plant food.

CHAPTER 5

Effect on Human Health

'Right feeding is the biggest single factor in good health—but the food must be *right in quality* as well as quantity.'

These words are taken from the *Daily Mail*, written by the Radio Doctor, a well-known voice on the air.

'Good health'—the feeling of wholeness, not the negative: 'I don't feel ill', but the positive, radiant, good health affecting body, mind and spirit.

Modern statistics show how rare it is.

From America: 'More than 4,000,000—i.e. one-third of its young draftees—were rejected, as physically or mentally unfit': and again: '95 per cent of Americans need some dental treatment'.[1]

The Peckham Health Centre, known as the 'Peckham Experiment', published some startling facts.[2] The Health Centre was a family club, under the supervision of *medical and biological* experts. The conditions of membership included a periodic *'health'* overhaul of the entire family, with a service for subsequent consultation and advice. Social and recreative activities formed an integral part of the scheme. The members were a cross-cut section of the community, and therefore a fair sample of the national health.

The statistics are startling:

[1] From *Organic Gardening*—Rodale Press, U.S.A.
[2] From *The Peckham Experiment*, by I. Pearse and L. Crocker. Published by George Allen and Unwin.

Effect on Human Health

Out of 500 families examined only 9 per cent of the individuals were 'healthy', i.e. 'without disorder'.

Out of a second list of 500, taken at random from a total of 1,206 families examined, only 10 per cent were healthy.

Pretty grim figures! What is wrong?

In *The Labouring Earth*[1] Mr. Alma Baker pointed out that general ill health was not confined to man. It is prevalent in domestic animals, and cultivated plants. If men were ill, and stock were healthy, or if animals were diseased and plants had good health, there would be no common ground for judgement. But as all three lack good health, there must be a common cause. Unhesitatingly he states that the common cause is the soil.

Again it is common sense! A devitalized soil cannot produce vital plants, and as the plant is the foundation of *all* food, whether animal or vegetable, if the *plant* is deficient in vitality *all* suffer alike.

Vitality is the one thing that man cannot give, clever chemist as he is.

There are many synthetic foods on the market today. Chemically speaking they may be perfect, but—I wonder! Have they life? Vitality? If not, they cannot give it. To my mind, there is only one way of testing food, and that is testing it on *life*. It can't be judged by the chemist's test-tube, or even by the *immediate* response of the human body; it may act as a dope, or a stimulant. Its *feeding* properties can only be judged by its effect on living entities, viz.: the white rats and other animals of the biologist's laboratory and a long-term test on human beings.

If such a test were made obligatory for all synthetic food—yes, and all artificially-fed vegetables—the safety of human health would be better guarded.

[1] *The Labouring Earth*, by C. Alma Baker (Heath Cranton).

Effect on Human Health

In *Your Daily Bread*,[1] a delightful book, full of wisdom, knowledge, stories, and a keen sense of humour, the author says: 'Why cannot man leave good food alone? It seems impertinence on his part to think he can improve on the wonderfully intricate and involved designs of Nature by processing, bleaching, refining, de-vitaminizing, by taking live things out and putting dead things back, most of all by separating the wholeness of foods. We are finding only too surely that this interference brings sooner or later its own penalties. In fact, it has been said that neglect or contempt of natural laws is the sole cause of all our misfortunes.'

One thing is certain. Nature is swift to respond or retaliate as man keeps or breaks her vital laws. Bread has not been called the 'Staff of Life' for nothing.

Surely in the name of common sense and national health, whole-meal bread will be obligatory in the future.

Add to devitalized plants and denatured food, the long list of poison sprays used as insecticides and fungicides.

Again it is common sense. If a poisonous remedy is strong enough to kill a pest, it is strong enough to harm a human being: not kill, but undermine his health: to say nothing of the harm done to the myriads of unseen soil workers, for no poison is wise enough to kill the pest, and spare the friend. All perish alike, to the detriment of the fertility of the soil. Feed the plant with natural humus, and it will give, as it was meant to give, full vitality to mankind.

Now for the other side of the picture. Are there any definite examples of improved health arising from fertile soil? There are—plenty. Examples have been given in *The Living Soil*, by E. Balfour, the most outstanding and constructive book on the subject. Here are a few others:

A well-known landowner in Surrey adopted a system of composting, in place of artificial manures. The produce

[1] By Doris Grant (Faber and Faber).

Effect on Human Health

was given to pigs and poultry, with the following results:
1. Mortality among new-born stock practically ceased.
2. General health of the stock improved.
3. A reduction of 10 per cent in the ration was obtained because of the *satisfying power* of the home-grown produce.

A large school with both day boys and boarders started the Indore method of composting for their vegetables, instead of using artificial manures. The results were both interesting and satisfactory.

Before the change-over, the school had suffered from epidemics of colds, measles, scarlet fever, and the like. After the adoption of compost-fed vegetables, illness was confined to sporadic cases, brought in from outside. In short, the disease resistance noticed in the plants was repeated in the humans who ate them. The 'common cause' is 'the fertile soil'.

In my own experience, several Q.R. compost users have written about the amazing improvement in the health of themselves and their households, since they started using the compost.

I have noticed the same improvement in my personal friends again and again—after a few weeks' visit.

Here is another story.

At the Anthroposophical farm in Holland, the produce was sent to customers direct, and by a specialized system of delivery and order. Costs were higher than the current market prices. In time a certain family demurred at the extra price, and returned to the market stall. After some time they came back to their old allegiance, saying they had had to pay so much in doctors' bills since eating the market stuff, that it more than counterbalanced the higher prices paid for vital food!

These are a few of the practical effects. But the tide of public opinion is slowly rising, and with the weight of medical

Effect on Human Health

statements, and growing conviction, the truth will have to be faced that indeed:

Right feeding is the biggest single factor in good health, but the food must be right *in quality* as well as quantity.

CHAPTER 6

The Activator

The Indore method mainly relies on animal manure for its activator; but we have to face the fact that the great majority of workers on the land cannot get manures.

Without an activator, the compost heap disintegrates slowly. Chemical activators break down the material, but destroy the working micro-organisms, and so do more harm than good.

I have urged the importance of retaining the natural heat of disintegration; in the pleasant atmosphere of subsequent moist warmth, the work of the micro-organisms proceeds apace. The quick breaking down of the fresh living plants, the disintegration of leaves, flowers, and stems, releases the life, the vitality, of the plants.

Life is eternal. It must go on.

Like all natural forces (water, electricity), it follows the lines of least resistance. With the enclosed heap, it cannot escape into the air. It turns back into the heap, vitalizing, energizing every part of it, all its internal activities.

Into this mass of pulsating life, we insert the herbal activator.[1] It holds the following ingredients, which contain

[1] The Q.R. methods must not be confused with those connected with Dr. Rudolf Steiner. He first advocated the use of the above-mentioned herbs in agriculture, as publicly stated by the Anthroposophical Society. The activator used in the Q.R. method is entirely different from the preparations made and used by the societies connected with Dr. Steiner's name.

This note is inserted to prevent any possible confusion.

The Activator

amongst them the chief elements needed by plant life:

Yarrow	*Chamomile*	*Dandelion*
Iron	Potash	Iron
Lime	Lime	Soda
Potash	Phosphorus	Potash
Soda	Sulphur	Phosphorus
Phosphorus		
Sulphur		
Nitrates		

Oak bark	*Valerian*	*Nettle*
Potash	Formic Acid	Oil
Lime	Acetic Acid	Formic Acid
		Ammonia
		Carbonic Acid
		Iron

Honey
Glucose

The herbs and honey are reduced to a fine, very sweet-smelling powder.[1]

The strength of the dose for treating the heap is: 1 grain (weight) to 1 pint of rain-water. (One grain will cover a sixpenny or an American cent piece.)

The grain is made up of seven ingredients, i.e. one-seventh of a grain of each!

How *can* it work? The answer is 'by radiation'.

Shake the bottle, as soon as the powder is dropped into it. The powder will rise, but nothing else will happen. Let the bottle stand for twenty-four hours. Shake it again. You will find a new activity, a bubbling, and a little foam on the surface. It has come to life. Smell it. The sweetness of the dry powder is in the liquid. Pour it into the heap, allowing about

[1] For formulae see Appendix, p. 81.

The Activator

three ounces to each hole; the holes are made twelve to twenty-four inches apart, and penetrate nearly to the base of the heap. Fill up the holes with soil and ram it down (this is important to avoid air spaces).

The whole process takes about ten minutes. When you open the heap, in four, six, eight or twelve weeks, according to the time of the year, you will find it evenly composted, and will discern in it the sweet smell of the herbal activator. How is this?

The water has released the living forces of the *elements* in the herbal powder. From the focal points at the bottom of the heap, these forces radiate upwards, and outwards; they diffuse yet more energy, more life, through the heap, and it is the energy and life of those *particular elements*, needed by plants, given *in plant form*, i.e. in the same rhythm of life that manifests in the vegetable kingdom.

But *why* bother with the powder, and the tiny dose? Why not use layers of the same weeds? Nettle and yarrow and the rest?

Why? Because the power in a large quantity would be so great that the radiations would pass *out* of the heap before releasing the forces they hold. The power when released would be great indeed, but it would develop in the upper air, and be lost to the heap.

With the minute dose, the elements within the radiations are able to develop and free their full power, *within the confines of the heap*. It is the same law that governs the fact that whereas a small dose of certain poisons will kill a man, a large dose will pass through him and leave him unscathed. It is the law of the 'power of the infinitely little'.

I cannot explain it further. All I *know* is that this minute quantity of herbs ripens, quickens, enriches the heaps without further turning, without interference, and after ten years of constant experience, I have *never* known it fail.

The Activator

THE HERBS

The list of herbs given is for the full formula, which is the one I use, as explained in Chapter 2.

Experiments have proved that any combination of herbs will work as an activator if they contain, between them, the chief elements needed by plant life provided they are used in homœopathic doses. This fact may be of real value, for people who are unable to find all the herbs in the full formulae.

Further, I believe it is better, if possible, to use plants indigenous to each country. There must be many plants in every land that contain the essential elements. Very few plants have been analysed for their constituents. I suggest that scientific herbalists in every country should investigate this question, and issue a list of native plants, and their essential constituents.

Ideally speaking, every farmer, every gardener, should be able to make his own herbal activator. The full formula is given in the Appendix (see p. 81). For those who have neither opportunity nor time to do so, the activator can be bought (see p. 83).

The following combinations have been tested with success.

Honey is always included; it is a powerful activator—*very* lively.

Nettle is an essential; it is the only plant I know containing carbonic acid and ammonia. Alternatives are:

1*st*	2*nd*	3*rd*
Yarrow	Chamomile	Chamomile
Nettle	Coltsfoot	Dandelion
Honey	Nettle	Nettle
	Honey	Honey

Of these, the first is the best.

The Activator

RADIATIONS

With the radio in every home, the mysterious power of radiation is generally accepted. With the new radio discoveries of the war and its marvellous developments, people are beginning to realize that its possibilities are unlimited, and its powers universal. Once this power is admitted, may it not explain many of Nature's secrets? Is it not even obvious, once we know where to look? Can we not discern it in the waves of scent that greet us from garden and hillside and wood? In the silent S O S of the ant struggling with a burden too great for her individual effort, and in the amazing hurrying response? Might it not explain the action of the trace elements? Boron, for instance, beneficial at one in ten million. And might not it explain this delightful story of the bees:

A radio station suffered, throughout the summer, from an epidemic of bee swarms! They came constantly and clung to the door of the building. Why? Surely by the fact that a wave-length used by the installation was the wave-length of bees. They tuned in, and literally arrived 'in their swarms'!

In my book *From Vegetable Waste to Fertile Soil*,[1] I wrote in 1940:

'When it comes to these fine radiations we are beyond the scope of material chemical analysis, and are within the sphere of physics—in the region of emanations, vibrations, waves, energy, forces of nature all recognized by modern science. Do horticulture and agriculture really shut themselves off from these realities? A man of science, a physicist, must soon arise who will investigate these proven facts, to find an explanation and open the scientific door to a pathway of discovery, a pathway that, judging from the results of

[1] Faber and Faber.

The Activator

practical experiments, will lead to better health of soil, of plants, and of mankind.'

A few weeks ago I read a book called *The Secret of Life* by Georges Lakhovsky, a Russian-born, naturalized French citizen, a scientist, an engineer physicist. His investigations on 'Radiations in relation to living beings' first appeared in 1923. His subsequent work with plants, animals and man, his theories and conclusions have been presented to the French Académie des Sciences by Professor d'Arsonval, spoken of as 'one of the greatest scientists of our time'. He also sponsors the book. It has been translated into five languages. The English version is the latest, published in 1939. It is amazingly interesting, and so simply written, so clear, that the layman can understand and follow it.

In the book Lakhovsky develops the theory that:

1. 'Every living being emits radiations.'
2. 'Every living cell is dependent on its nucleus, which is the centre of oscillations and gives off radiations'.

He states that the cell, essential organic unit in all living beings, is . . . an electro-magnetic resonator capable of emitting and absorbing radiations of a very high frequency.

To the question, What is life? he answers: '*It is the dynamic equilibrium of all cells, the harmony of multiple radiations which react upon one another.*'

He holds that all disease comes from the dis-equilibrium (unbalancing) of the vibrations of the oscillating circuit, i.e., the nucleus of the cell. This can be effected by the stronger vibrations of an invading cell, i.e. a microbe. Health, resistance, can be achieved by strengthening the natural vibrations of the weaker cell by outside interference. He links all vibrations with the Cosmic Rays, in which he says: 'Every frequency finds its counterpart.'[1]

[1] By a simple device Mr. Lakhovsky succeeded in filtering the cosmic rays and used the device to cure plants of tumorous growths.

The Activator

He speaks of '*the individual frequency of each cell*', and states further 'that *each group of cells* has its own frequency, with its own characteristic vibrations'.

Do we find here the scientific explanation of the radiations in the compost heap? I wonder.

The Q.R. Powder

During the past year I have received some interesting letters about the action of the Q.R. powder. The correspondents state their own theories on the subject. One is of such wide interest that I should like to pass it on.

The writer is a retired doctor. He believes the activity of the Q.R. is due to the work of enzymes, yeasts or moulds. He relates the following story:

Before the First World War he went to Copenhagen to visit a famous brewery. The chief brewer told him that he, personally, had used, with success, over 2,000 different sorts of yeast for the brewing of beer, and there were many others he had not yet tried. On his return journey to England the doctor travelled with the head brewer of the Brewery Union of South Africa who told him that he got his yeast once a year from the Copenhagen firm. 'How do you manage transport?' asked the doctor. 'Very simply,' said the other, 'it comes by letter post. The chief brewer takes a sterile knife,

He later developed this device into an instrument known as the Multiple Wave Oscillator. This instrument, based on his theory, has been tested and used by the leading medical faculties on the Continent.

Since its inception in 1931, it has been installed in hospitals in France, Italy, Germany and Sweden. The book has many illustrations of the results of its use, especially in reference to cancer.

Georges Lakhovsky has been awarded the red ribbon of the Legion of Honour for his services during the war.

The book is published by William Heinemann (Medical Book Dept.).

Since writing this I have learnt that G. Lakhovsky escaped from France, but died in New York in 1943: a terrible loss to the world of science.

The Activator

scrapes off a little of the culture, wipes it on a piece of sterile paper, folds it, puts it in an envelope and posts it to me. I scrape off the culture with a sterile knife, stand it in a jar of sugar and water at a certain temperature, and from that I get enough yeast to brew millions of gallons of beer'. 'Why then do you have to send back annually for a fresh supply?' 'Because' (and *this* is the interesting part), 'because our Brewery is in the middle of the vine growing district of South Africa. The bloom on grapes is a yeast, it contaminates my yeast, renders it useless and I have to replace it.'

The doctor suggests that the Q.R. powder may contain the spores of a yeast or enzyme which stimulates the fermentation of the heap. Enzyme, yeast and mould are closely allied. The story is interesting and suggests a new line of approach.

Another opinion comes from a Dutch soil expert who is working on the devastated soil in Holland, and who has been experimenting with Q.R. for some months. He wrote that in his opinion 'The radiations of the Q.R. activator are strong stimulus to the life and development (procreation) of the micro-organisms in the heap, and it is for that reason you can only apply it in infinitesimal quantities. The strong radiations of greater quantities hamper, or perhaps even destroy that life.'

Of the compost itself he says, 'The two predominant qualities of the Q.R. Compost are:

'Well-*made* it is a food extract for plants.

'Well-*treated* it is a rich store-house of micro-organic life.'

His remarks are founded on practical experiments and practical results, and he is continuing the tests.

It is very interesting to notice how modern science is approaching the theory of Radio-biology. Professor Libby of Chicago University stated, 'That there is more radio-active carbon to be found in human beings, animals and plants than ever physicists are likely to make by transmuta-

The Activator

tion in the laboratory' (from *Mother Earth*, Autumn 1947, quoted from *Science Today*), and Dr. E. Pfeiffer in *Organic Gardening* (Rodale Press) writes, 'There is more radio-activity in the leaf of the duckweed than there is in the water in which it lives.'

This all points to a whole realm of new knowledge, to a vast unharnessed natural force subject only to Nature's laws, universal, ubiquitous, and I believe in time it will be found to be the ultimate power behind the 'mysteries' of Nature. A force which man can distort to his own undoing, but with which he can co-operate by close observation, an open mind and patient research.

CHAPTER 7

The Conviction

In this chapter we come to the question: What is the conviction behind it all?

I will quote from my first book, *From Vegetable Waste to Fertile Soil*.

'The method is based on two main convictions:

'1. That all growth is the effect of the interplay of living forces—not the result of automatic chemical change. That these forces pass through soil, permeate atmosphere, are carried by the elements, and are behind the mystery, the vitality of plant growth.

'With quick and controlled disintegration, these living forces are released and radiate into the heap. There they work in a vast co-operation with fungi, bacteria, earthworms, and other soil workers, and are returned to the earth strengthened by the herbal essences, ready for use once again for plants and in the same rhythm of life as the plants themselves.

'2. There is life throughout the universe—life, manifesting at a different rhythm in each of the four kingdoms.'

Those convictions remain unchanged but strengthened.

One life manifesting in each of the four kingdoms, but at a different rhythm.

There *is* life in a stone, otherwise it would fall apart and become dust. But a stone cannot grow; the life within it must pulse at a different rate to life in the vegetable kingdom

The Conviction

in which plants even in the lowest forms, such as lichens, grow, and die, to become humus—i.e. vegetable manure.

There are many degrees, many links between the four great kindgoms, but always there are definite differences.

The plant grows; it has vitality, that same vitality which is the basis of all food; but it is restricted. Life gives it much, but it cannot move, it can express no emotion, it has little free choice, it must take what it finds within reach of its roots.

The animal kingdom is a step higher, life is fuller, manifesting in a more complicated way, with larger possibilities, swifter action, greater intelligence. The rhythm of life is quicker.

Then comes man—with all his potentialities of body, mind, and spirit, of service and sacrifice, of invention and thought, often a battleground of conflicting desires, emotions, and aspirations.

There is a great gap between the plant and man; yet man is linked to the plant by the *vitality* that comes to him in food, either direct through vegetables, or via the animal; but to feed the plant *directly*, I repeat, for that is the operative word, to *feed* it *directly* with raw matter belonging to other kingdoms, either the mineral (salts) or animal (blood) is to introduce a different rate of life into its being, and thereby unbalance its own rhythm and impair its constitution. Feed it within the rhythm of its own kingdom and all will be well.

The basis of animal manures is plants in an advanced state of decomposition. It is the strong animal juices, digestive juices present in *fresh* manure, that burn plants and attract pests. Old manure, or composted manure, is one of the finest forms of humus, and plants feed mainly on the humus that they find in the soil.

Yet there *is* Life in all that exists, and *what is life*?

The Conviction

Do we not find the answer in the words of the Eastern sage as he writes of:

'The Divinity that sleeps in the stones, stirs in the plants, wakes in the animals, is conscious alone in man.'

The breath of God in all that is.

That is Life.

Appendices

1. TABLE I. BUILDING THE GARDEN HEAP

Foundation	*Good Drainage*
Bin	Wood, if possible, or bricks, or turves, grass down
Air	*Essential:* spaces between boards, or holes in sides of bin
Size	From 18 in. × 18 in. × 2 ft. high to 6 ft. × 6 ft. × 3 ft. high
Time of building	Up to two months
Method	In 4 in. layers, kept flat: alternate tough stems with soft green
Materials	*All* vegetable waste 4 in.: a dusting of lime, three times. A few spadefuls of soil. An occasional 2 in. of manure. A final cover of 4 in. soil
Protection	Keep a sack on last layer ALWAYS
Shelter	Tilted corrugated iron sheet, or a stretched canvas cover, against rain
Treatment	Inject activator when heap is full and firm: 1 pint per 6 ft. area or use horizontal method (see p. 40).

TABLE II. BUILDING THE STRAW HEAP

Size	6 ft. wide × 8 ft. long × 6 ft. high, in sections

Appendices

Layers	4 in. *wet* straw, 2 in. green matter, or 2 in. manure, or a scatter of soil
Method	Build in open, like a rick, with straight sides, ridged top
Shape	Long clamp; build in sections
Cover	Soil
Treatment	Activator: 1 pint per 6 ft. area

TABLE III. MANURE TUBS

1.	2.
Sink tub to within 6 in. of its rim	Sink tub
Fill with *fresh* manure	Fill with *old* cow-pats
Treat 3 oz. activator	Treat 3 oz. activator
Cover with lid	Cover with lid

TABLE IV. LEAF HEAP

Stack in wire enclosure
Scatter soil every foot
Turn out in six months
Treat with activator—1 pint per 6 ft. area

TABLE OF FAILURES, CAUSES AND REMEDIES

After ten years of constant work there has not been one failure in this garden. A few have been reported, and invariably one of the following causes had been traced:

1. Loss of heat.
2. No aeration.
3. Misuse of activator.
4. Opening the heap before ripe.

Appendices

Cause	Reason	Result
1. Loss of heat	1. No sacking laid on top layer	Dried out (*a*)
	2. No shelter	Rain drenched heap (*b*)
2. No aeration	1. No air spaces in walls of bin	Putrefaction (*b*)
	2. Too heavy treading caused tight packing	Putrefaction (*b*)
3. Misuse of activator	1. Not used	Heap not disintegrated (*c*)
	2. Kept over a year	
4. Opening before ripe	Opening by a fixed date	Not yet decomposed (*d*)

Remedies

(*a*) Pour one gallon manure or compost water over it.
(*b*) Fork into pile, let sun and air get to it, then wet with manure water.
(*c*) Remake with fresh green layers and treat.
(*d*) Wait—a week or more.

2. FORMULAE FOR HERBAL POWDER

Material[1]

1. Wild Chamomile (*Matricaria chamomilla*).
2. Common Dandelion (*Taraxacum officinale*).
3. Common Valerian (*Valeriana officinalis*).
4. Yarrow (*Achillea millefolium*).
5. Stinging Nettle (*Urtica dioica*).
6. Oak bark (*Quercus robur*).
7. Pure Honey.

---

[1] 1956. The addition of a small proportion of dried selected seaweed marks a further improvement in Q.R.

Appendices

Method

Gather flowers and leaves before midday. Dry as soon as possible with slow heat, i.e. on hot water pipes or under a raised stove. When tinder-dry crush and pass the herbs through a fine wire sieve (a kitchen sieve) or a bag of book muslin. Keep each of the herbs separate.

Oak bark: Use the outside rough bark, grind or rasp it to a powder, pass it through the sieve.

Honey: Rub one drop of honey into one dram of sugar of milk till the honey is completely absorbed. (Sugar of milk is a pure product used to feed babies, obtainable at any chemist's or chain stores.)

For Stock Mixture

Take equal parts (say a level teaspoonful) of each of the ingredients, mix them thoroughly, keep them in a covered jar.

For Use

Stir again to ensure an even mixture, and liquefy as follows:

To Liquefy: Mix as much of the powder as will cover a sixpence, or one cent piece (approx. one grain) with one pint (20 oz.) rain-water. *Shake well.* Let it stand for twenty-four hours before using. It will keep good for about three weeks. Shake thoroughly before use.

Note: The two essential ingredients are yarrow and nettle. The others are used because of their prophylactic qualities; if unobtainable, any one of them may be omitted.

For treating the heap see page 40.

Please note, that both methods are very elastic, and allow much latitude every way: *exact* measurements and *exact* doses are not essential, and the amounts must be fitted in to the size of the heaps.

Appendices

An 18 in, square area would take three holes, arranged (a) · . ·

A 4 ft. × 3 ft. area, would take five holes, arranged (b) ⋮ ⋮

A 6 ft. × 6 ft. area, would take seven holes, arranged (c) ⋮ ⋮ ⋮

3. PRICE OF POWDER ACTIVATOR

For those who cannot make their own activator, the powder is available at the following prices:

Packets 12p and 32p; Carton £1.45

(all inclusive of postage)

direct from:

Chase Compost Seeds Ltd., Benhall, Saxmundham, Suffolk. The powder is also obtainable at most garden supply stores.

Direction for use:

To Liquefy: Mix a small teaspoonful in a pint of rain water, shake well. Let it stand for four hours. It will keep about three weeks or more; when it smells sour—discard it.

Important: Always shake the bottle *well* before use. A convenient way of carrying the solution is in a pint beer bottle.

To use: follow horizontal method described on page 40.

4. HERBS—AND WHERE TO FIND THEM

Wild Chamomile:

Flower	A delicate daisy with a protruding yellow cone. This cone is *hollow inside like a horn*. This is an important distinguishing mark; in all the other species the cone is fleshy
Height	From one to two feet. Branched in growth
Leaves	Bright green, delicate, fern-like
Grows	In cornfields, roadsides, rough fields: will grow easily from seed in light soil
Season	June to September

Appendices

Common Dandelion:

Flower	Bright yellow, large many-petalled daisy
Leaves	Flat, long-toothed, deep green
Flower Stems	Bare, hollow, with milky juice
Grows	Everywhere. A very common weed
Season	From April to September

Common Valerian:

Flower	Pale mauve, small, grow in a many-flowered flattened cluster
Height	From one to four feet, branched stem
Leaves	In pairs up the stem; they are united at the base, and are cut in segments opposite each other. The root leaves grow singly, but are cut in segments
Grows	On damp banks, hillsides, and in ditches
Season	June to August

Common Yarrow:

Flower	Flat heads of daisy-like flowers, white or pale pink.
Height	From one to two feet, branched stem
Leaves	Finely dissected, alternately arranged on stem
Grows	On waysides, fields, dry banks: very common

Stinging Nettle: (this must *not* be confused with the dead-nettle family, *Labiatae*)

Flower	Green, hanging in clusters from the axle of the leaves
Height	From two to five feet
Leaves	Heart-shaped, toothed, dark green. The whole plant is covered with stiff hairs, that sting when touched
Grows	On waste ground, roadside, in hedges, everywhere

Appendices

5. ALTERNATIVE PLANTS AND THEIR CONSTITUENTS

Hollyhock
(*Althea rosea*)
Iron, Soda
Potash, Phosphorus
Lime, Sulphur

Strawberry
(*Fragaria vesca*)
Lime
Soda
Phosphorus

Marigold
(*Calendula afficinalis*)
Lime
Sulphur

Elder
(*Sambucus nigra*)
Iron
Potash
Soda

Walnut
(*Juglans nigra*)
Iron, Phosphorus
Potash, Sulphur

Yellow Dock
(*Rumex crispus*)
Iron
Sulphur

Sage
(*Salvia officinalis*)
Potash
Lime
Soda

Chicory
(*Cichorium intybus*)
Iron, Potash
Lime, Soda

6. SOME USEFUL HINTS

To increase heat:
   Bruise and wet some nettles, put a layer in the heap.

To clean Fruit Trees:
   Take two handfuls of dried horsetail (*Equisetum arvense*) and simmer in a gallon of water for twenty minutes. Use one pint in one gallon water and spray, and in bad cases scrub your fruit trees. It is a wonderful cleanser. Strain and bottle any liquid left over.

Appendices

To kill American Blight:

Use the horsetail liquid at full strength, dab or brush it on to the colonies.

To kill Aphis:

Dry and powder bracken leaves. Soak one dram (weight) in six ounces of water. Let it stand for twenty-four hours, strain and bottle. Use one dram in one gallon rain-water. It should tinge the water green. Spray, or better still, wash the leaves attacked by aphis.

Appendices

7. Plan for a Small Bin

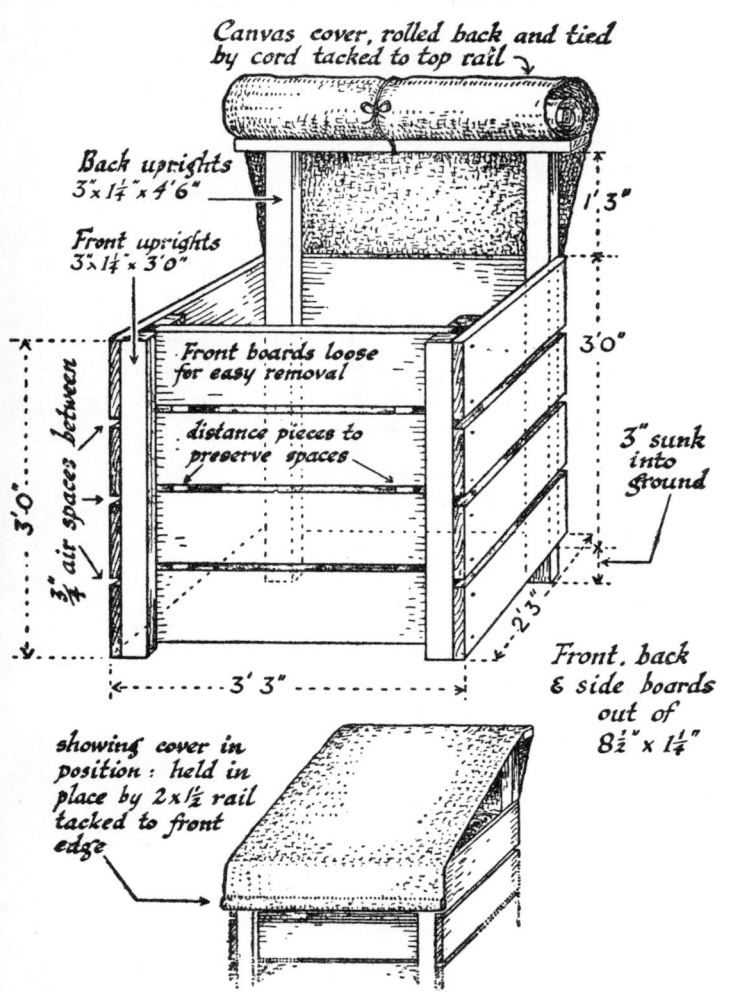

Canvas cover, rolled back and tied by cord tacked to top rail

Back uprights 3" × 1¼" × 4'6"

Front uprights 3" × 1¼" × 3'0"

1'3"

Front boards loose for easy removal

3'0"

distance pieces to preserve spaces

3'0"

¾" air spaces between

3" *sunk into ground*

2'3"

3'3"

Front, back & side boards out of 8½" × 1¼"

showing cover in position: held in place by 2 × 1½ rail tacked to front edge

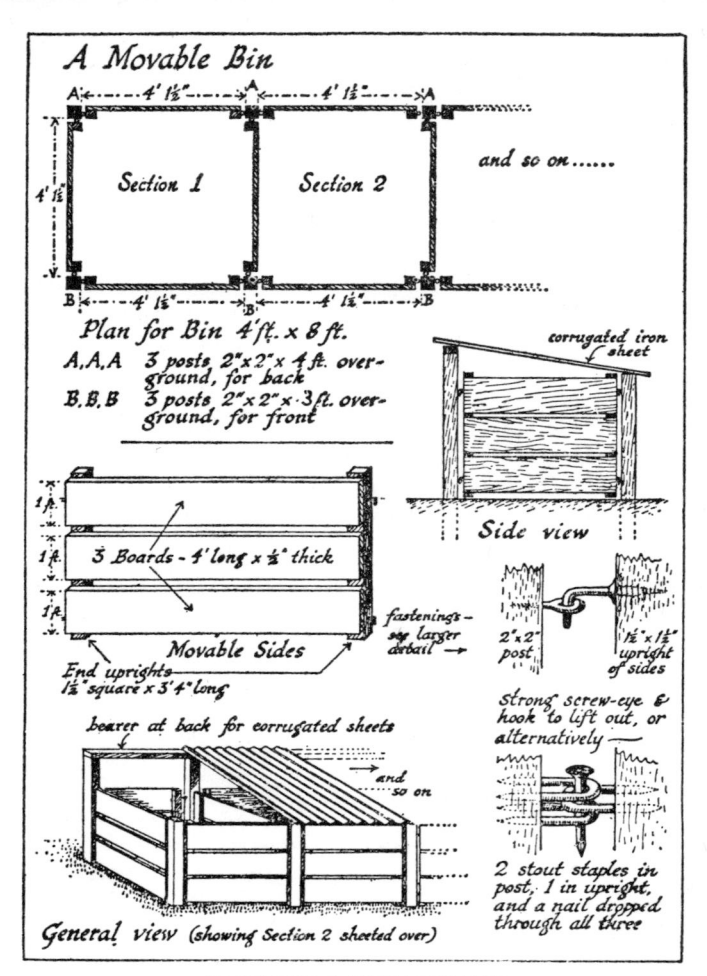

PLAN FOR A MOVABLE BIN

Method of Erection

The posts can be permanent if desired. The sides are made light and easily moved and fixed. The first heap can be completed in section 1 and the other sections added on as required. Note that of the 4 ft. 1½ in. between the posts, 1½ in. is to allow space for the hook and eye or staple fastening.

Appendices

8. Plan for a Movable Bin

The movable bin has been designed to meet the need of a bin which is easy to erect, and dismember, and moreover which is light, efficient, and capable of infinite expansion. It would be useful in outlying places, in farms or gardens.

The general layout is a series of square sections, with light movable sides which are hooked on to four equidistant corner posts. Each section is independent but additional ones can be added to any length required, i.e. a 4 ft. section could be expanded to 4 ft by 8 ft., 12 ft., 16 ft., and so on. The corner posts could be permanent if desired. They are firmly driven into the earth; the back ones are 4 ft. and the front ones 3 ft. above ground. This allows for the slope of the protecting cover. The sides are light panels made of ½ in. boards, 1 ft. wide, nailed at each end to two upright spars, 1½ in. square and 3 ft. 4 in. in height. There should be a ½ in. space between the boards for aeration. They should not be *less* than 2 ft. nor *more* than 4 ft. long.

Note. A space of three quarters of an inch should be allowed each side between the panels and the stationary posts to allow for the fastenings.

The fastenings are strong, screw-in hooks and eyes. Two hooks on each side of the panel at top and bottom. The corresponding eyes in the stationary posts.

The sides can very easily be dropped into place and with the two fastenings remain rigid. For full details see plan, opposite.

Appendices

Here is a simple way of constructing a home-made 'straw and wire' bin. *Material.* A 12 ft. length of wire netting 6 ft. high and up to 1½ inch mesh; a bale of straw; and a short length of faggoting wire. *Method.* Fold the wire in half lengthways.

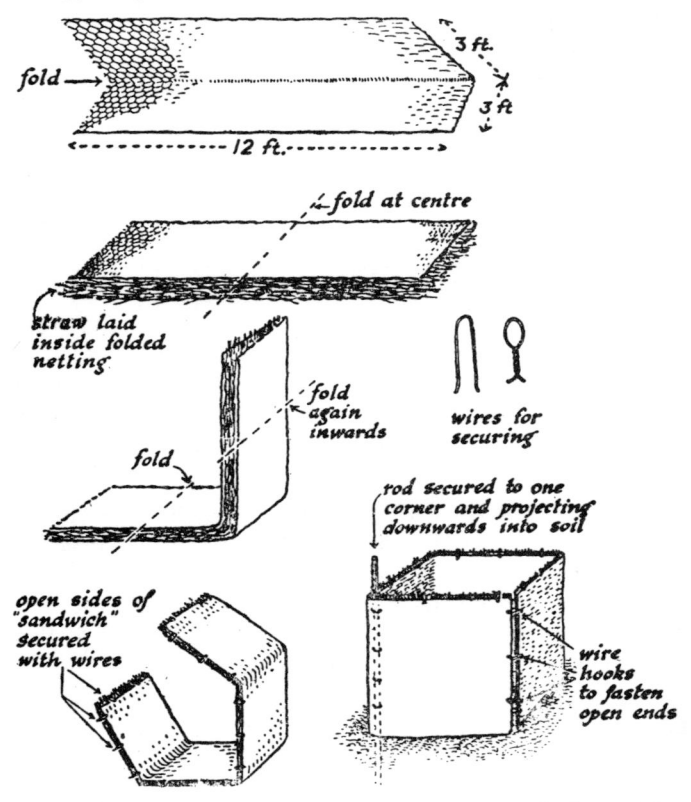

Tread the crease heavily. Pack the straw evenly in a 3 in. layer between the folded wire, like a straw sandwich. Fold in half. Tread the crease. Open and fold the ends to the centre. Tread the creases. Pick it up, it will go into a perfect square.

Appendices

Fasten the straw and wire with bits of faggoting wire, shaped like a hairpin, and twist the ends together. One iron rod through one end fixes the bin to the soil. The other end is closed with three wire hooks. An extra panel of wire and straw, supported by two stakes and a cross-bar, makes a roof This bin is successful and lasts a considerable time.

Sacks. It is worth asking a corn merchant if he has any old sacks to dispose of. Many which would protect a compost heap are sent away as 'junk'.

9. Tests at the Haughley Research Farm

One of the outstanding events of the year was a letter from Lady Eve Balfour, asking if I would submit the Q.R. Activator to a scientific test at the Haughley Research Farm. I was overjoyed! It was the fulfilment of a wish of many years' standing.

The test was started in July, on the normal farm heaps, i.e. built on the Indore Method. The supervisors were Lady Eve Balfour, Dr. J. W. Scharff, and myself. Students at the Haughley Farm were present. The following report was received in December 1946:

July 15th and 16th. Made *two double sections* of compost, each double section measuring 10 ft. × 10 ft. × 4½ft.

Material. Long muck from cattle yard, a little soil and chalk, then layers of nettles put between loads as heaps were built. Heaps ventilated with three vertical vents to each 10 ft × 4½ ft. section. Heaps covered with layer of soil, otherwise no covering.

July 19th. Third double section made as above. (All sections were surrounded and separated by walls made with bales of straw.)

July 21st. A (*First double section*) treated with Q.R. activator poured into holes at two-foot intervals and afterwards filled up with soil

Appendices

B (*Second double section*) holes were made and filled with soil as in A, but no Q.R. herbal infusion inserted.

August 28th. Turned third section (C).

October 7th. A and B turned (first time). C turned (second time).

Short summary from the interim report by Dr. J. W. Scharff on the inspection and turning, October 7th:

A (*treated heap*). Humus development nearly complete, friable, sweet smelling, condition corresponding to what might be expected at the second turn.

B (*untreated heap*). Humus development incomplete, much of it tough, unrotted, with patches smelling of ammonia.

'There is unanimous agreement on the considerably greater breakdown of organic matter and increase of humus formation in the treated heap. From the data already gathered it would appear evident, that at the very least, the inoculation of Indore compost heaps with Q.R. Herbal infusion and soil will enable a saving to be made of at least one turn.'

Continuation of Lady Eve Balfour's Report:

December 4th. A, B and C were all spread on the land. The treated heap (A) was the only one that was still warm. Half-sack samples were taken from the centre of each heap. (Note. All heaps suffered from excessive rainfall.)

December 7th to 14th. One shovelful from each sample, air-dried in a warm room and crumbled with fingers and rubbing between palms.

December 15th. Each of the three dried samples was stuffed as tightly as possible into a ¼-lb. jam jar. Each jar emptied in turn and put through a 1-in. sieve.

Appendices

RESULTS IN PARTS BY VOLUME

Sample	Passing Sieve	Not passing	Total
A. Q.R. treated; earth-filled holes; one turn	14	3	17
B. Untreated; earth-filled holes; one turn	12	5	17
C. Standard; two turns	12¼	4¾	17

IN PARTS BY WEIGHT

A. As above	21	3	24
B. As above	20	6	26
C. As above	19	6	25

This report was signed by Lady Eve Balfour.

The compost was used for the 1947 sugar beet crops; the following report is quoted from *Mother Earth*, Spring 1948:

'The total acreage sown was 6 acres, but as 1 acre was very poor indeed, calculations have been made on a basis of 5½ acres. All yields were low in 1947, owing to the late spring followed by drought, but the half-acre which received the compost treated with Q.R. yielded at the rate of 11 tons per acre, twice the tonnage of the average of the whole field, and approximately 2 tons per acre more than the next best half-acre' (Report by Lady Eve Balfour).

1947 Test

In 1947 a 'no turning' test gave interesting results. All the heaps were built on the Indore method in the open. The test was based on the percentage passing through a ¼-in. sieve.

Material	Treatment	Age	Result
A Indore Compost	Twice turned	5 months	97·73
B Indore Compost	Unturned	8 months	75·22

Appendices

C Indore Compost	Unturned, treated by sprinkling Q.R. solution between layers	8 months	92·57
D Indore Compost	Unturned, injected vertically with Q.R. solution	7½ months	86·7

'From this it would appear that for the farmer who has not the time to turn his heaps, the next best method for producing the greatest amount of breakdown in a given time is to sprinkle with Q.R. solution between the layers' (Lady Eve Balfour's report, *Mother Earth*, Spring 1948).

10. *The Soil Association*

The Soil Association is a registered Charity in the United Kingdom, and was founded in 1946. It has a membership of some 5,000 in 63 countries. It is best described by quoting from one of its Journal Editorials:—

"The Soil Association was founded by a group of people from many different walks of life, who believe that the right approach to a better understanding of health is the positive one of promoting vitality rather than the negative one of preventing disease.

To promote vitality, we must understand its origin. Our bodies are built out of the food we eat, and, directly or indirectly, that food comes from the soil. Science has taught us that even a single spoonful of soil contains millions of living creatures which thrive amongst the naturally decaying residues of plants and animals and in their turn support and nourish new growth.

As soon as it is realised that the soil itself is a segment of the cycle of life, the fundamental importance of maintaining its vitality becomes apparent. Health is seen as a wholeness and nutrition as a continuous flow of substances from the

Appendices

soil and back to the soil. This is what is meant by the law of return. There is reason to believe that the interruption of this cycle at any point leads, slowly but surely, to disease in plant, animal and man, and in body, mind and spirit.

This we seek to investigate through the science of ecology, which is the study of the relationship of living things with each other and with their environment. Scientists have discovered a great deal about the physics and chemistry of the soil, but it is mainly to a few pioneers in organic cultivation that we owe the comparatively little that has yet been learned about soil ecology. The Soil Association believes that it has an important contribution to make in this vital field of knowledge.

The Soil Association is unique in conception and purpose because it seeks to bring together a great variety of men and women ranging from the trained scientist, investigating the relationship between the soil and vitality, to the ordinary citizen who sees in the organic attitude of mind a clue to health and a full life. The Association therefore has a wide task."

The Soil Association needs the support of public opinion if it is to make its voice heard. It has 33 County Groups in the United Kingdom who play an active part in furthering its aims. The Association publishes its own Journal and a monthly newspaper, which are sent free to all members. For further details, apply to:

>The General Secretary,
>The Soil Association,
>Walnut Tree Manor,
>Haughley,
>Stowmarket,
>Suffolk,
>England.

Bibliography

BALFOUR, E. B., *The Living Soil*, Faber & Faber.
BAKER, ALMA, *The Labouring Earth*, Heath Cranton.
BELL, JOHN, *Nature's Remedies*, Pitman.
BILLINGTON, F. H. AND EASEY, BEN, *Compost for Garden-plot or 1,000-acre farm*, Faber & Faber.
BROWN, S. A., *Compost Flower Growing*, W. Foulsham & Co.
CARSON, RACHEL, *Silent Spring*, Hamish Hamilton, Penguin.
COMMONER, PROFESSOR BARRY, *Science and Survival*, Gollancz.
DUDDINGTON, C. L., *The Friendly Fungi*, Faber & Faber.
EASEY, BEN, *Practical Organic Gardening*, Faber & Faber.
FURNER, B. G., F.R.H.S., *Compost Vegetable Growing*, W. Foulsham & Co.
GRANT, DORIS, *Housewives Beware*, Faber & Faber.
GRANT, DORIS, *Your Daily Bread*, Faber & Faber.
GRIEVE AND LEYELL, *A Modern Herbal*, Jonathan Cape.
LAKHOVSKY, G., *The Secret of Life*, W. Heinemann.
PEARSE, T., *The Peckham Experiment*, G. Allen & Unwin.
PICTON, DR. L., *Thoughts on Feeding*, Faber and Faber.
RUSSELL, SIR JOHN, *The World of the Soil*, Collins, Fontana Library.
SEIFERT, ALWIN, *Compost*, Faber & Faber.
WILLIAMS, S. R., *Compost Fruit Growing*, W. Foulsham & Co.

Index

American Blight, 84
animal digestive juices, 41
Anthroposophical farm, Holland, 10, 22, 63
Aphis, 84

Baker, Alma: *The Labouring Earth*, 61
Balfour, Lady Eve, 89-92; *The Living soil*, 62
beans, 52
Billington, F. H.: *Compost*, 20-1
bin, 33-5; construction, 25, 85-9; foundation, 35; movable, 86-7; plan, 85; protection, 34-5; size, 34
Bio-dynamic Farms, 22
Bio-dynamic method, 21, 23
bracken, 44

cells, 70-1
chamomile, 66, 68, 81
charcoal, 35
Chase Compost Seeds Ltd., 81
chemical activator, 65
chicory, 83
chipwood bedding, 49
coltsfoot, 68
cow-dung, 43
Crocker, L.: see Pearse, I.

dandelion, 66, 68, 82
d'Arsonval, Professor, 70
disease, 19, 31, 63; plants, 15, 32, 56-7
disintegration, 26, 33, 39-40, 45-6, 65, 74-5.
drainage, 44

Easterbrook, L. F., 29

elder, 83
elements, 29, 65-6
erosion, 17
experiments and tests, 22-32, 89-92

failures, causes and remedies, 78-9
fertility, 17
fertilizers, chemical, 17-8, 59
flavour, crops, 14-5
food, animal, 63; synthetic, 61-2; whole, 10
From Vegetable Waste to Fertile Soil, 69-70, 74
fruit, 15, 52, 54-5, 83

garden, 51-9
Grant, Doris: *Your Daily Bread*, 62
growth, 74

Haughley Research Farm, 89-92
health, 11, 15, 19, 60-4, 70; statistics, 60-1
heap, 31, 33-50; construction, 26, 35-9; farm, 39, 44-5; foundation, 35; garden, 77; leaf, 78; materials, 35-7; open, 46; ripening, 39; straw, 44; tests, 25, 89-92; treading, 47
heat, 37, 45-6, 65, 83
herbal activator, 21, 30, 38, 41, 47-8, 65-73; homœopathic dose, 24, 26; powder, 32, 38, 66-7, 71-3, 79-81; strength, 24, 66-7
herbs, 24, 29-31, 68, 81-2
hollyhock, 83
honey, 66, 68

Index

Howard, Sir Albert, 21
humus, 17-18, 40, 42, 48, 62

Indore method, 21, 63, 65, **89**, 91-2
insecticides, 19, 62

King, Mr., 46
kitchen refuse, 35-6

Lakhovsky, Georges: *The Secret of Life*, 70-1
leaves, Autumn, 40; heap, 78
Libby, Professor, 72
life, 74-5
lime, 37, 59

manure, 40-3, 75; liquid, 42, 47; synthetic, 16, 19; tub, 42-3, 78; water, 40, 48, 52-3
marigold, 83
marketing, 58
microbes, 33, 70
micro-organisms, 65, 67
minerals, 29

nettle, 30-1, 44, 66, 68, 80, 82

oak bark, 66
oats, 59

Pearse, I.: *The Peckham Experiment*, 60
peas, 52
Peckham Health Centre, 19, 60
pest resistance, 56
Pfeiffer, Dr. E., 22; *Organic Gardening*, 73
plants, 24; quality, 57-9; weather resistance, 55-6
pollution, 10
poultry manure, 14, 43

rabbit manure, 14, 43
radiation, 30-1, 66-7, 69-71
radio-biology, 72
rain, effect on heap, 34-5

sacking, 37, 89
sage, 83
Scharff, Dr. J. W., 89-90
seaweed, 50
smell, 28, 38-9, 40, 47, **51**, **57**, 66-7
Soil Association, 92-3
soil, 19, 51, 61; use of compost in, 51-2
Steiner, Dr. Rudolf, 13-14, **21-2**, 65
Steiner method: *see* Bio-dynamic method
storage, 40
straw, 44, 48; heap, construction 77-8
strawberry, 54-5, 83

texture, 46
tomatoes, 52
toxic sprays, 10
trees, 52

valerian, 66, 82
vegetable waste, 14, 35, 41
vegetables, 52
vitality, 15, 31, 59, 60-1, 74-5

walnut, 83
waterweed, 48-9
weeds, 36-7
wheat, 58

yarrow, 30-1, 66, 68, 80, 82
yeast, 71-2
yellow dock, 83